ALSO IN THE EXPONENTIAL MINDSET SERIES

Abundance: The Future Is Better Than You Think

BOLD: How to Go Big, Create Wealth, and Impact the World

THE FUTURE IS FASTER THAN YOU THINK

HOW CONVERGING TECHNOLOGIES ARE TRANSFORMING BUSINESS, INDUSTRIES, AND OUR LIVES

PETER H. DIAMANDIS
AND STEVEN KOTLER

SIMON & SCHUSTER

New York London Toronto Sydney New Delhi

Simon & Schuster
1230 Avenue of the Americas
New York, NY 10020

First Simon & Schuster hardcover edition January 2020

Manufactured in the United States of America

1 3 5 7 9 10 8 6 4 2

Library of Congress Cataloging-in-Publication Data

Names: Diamandis, Peter H., author. | Kotler, Steven, 1967– author.
Title: The future is faster than you think : how converging technologies are transforming business, industries, and our lives / Peter H. Diamandis and Steven Kotler.
Description: New York : Simon & Schuster, 2020. | Includes bibliographical references and index.
Identifiers: LCCN 2019044600 | ISBN 9781982109660 (hardcover) | ISBN 9781982109677 (trade paperback) | ISBN 9781982109684 (ebook)
Subjects: LCSH: Technological innovations—Economic aspects. | Technological innovations—Forecasting. | Technology—Social aspects. | Convergence (Economics)
Classification: LCC HC79.T4 D4895 2020 | DDC 338/.064—dc23
LC record available at https://lccn.loc.gov/2019044600

ISBN 978-1-9821-0966-0
ISBN 978-1-9821-0968-4 (ebook)

I dedicate this book to all who have mentored and coached me during my life: Harry P. Diamandis, Tula Diamandis, Frank Price, David C. Webb, Paul E. Gray, David E. Wine, Gregg E. Maryniak, Ayn Rand, Art Dula, Robert Heinlein, Byron K. Lichtenberg, Sylvia Earle, Gerard K. O'Neill, Arthur C. Clarke, John T. Chirban, Laurence R. Young, Martine Rothblatt, Charles Lindbergh, Tom Velez, Stuart O. Witt, S. Pete Worden, Robert K. Weiss, Alfred H. Kerth, Burt Rutan, Anousheh Ansari, Tony Robbins, Ray Kurzweil, and Dan Sullivan.

—Peter

This one is for the late Joe Lefler and the crew from Pandora's Box. Thanks for so much magic. Thanks for believing in me long before anyone else did. Thanks for Derek Dingle's cockroach pass. Still miss you. Down the funny stairs.

—Steven

CONTENTS

FOREWORD

Your authors met in 1999. Steven was working on an article about Peter's organization, the XPRIZE, which was then focused on unlocking the space frontier. Peter was working on, well, unlocking the space frontier.

Very quickly we discovered a shared obsession with cutting-edge technology and the use of that technology for tackling seemingly impossible challenges. This overlap led to a great friendship and a multi-decade writing partnership, of which *The Future Is Faster Than You Think* is the latest installment. This is our third exploration of how technology can extend the bounds of possibility and transform the world. Technically, it's also the third book in "The Exponential Mindset Trilogy," a series that includes this work and our two prior works, *Abundance* and *BOLD*. You don't need to read these books before diving into this one, but a little context is helpful.

Abundance is a book about how accelerating technologies are demonetizing and democratizing access to food, water, and energy, making resources that were once scarce now abundant, and allowing individuals to tackle impossible global challenges such as hunger, poverty, and disease. In *BOLD*, we tell the story of a different impossible: how entrepreneurs have been harnessing these same technologies to build world-changing businesses in near record time, and providing a how-to playbook for anyone interested in doing the same.

In this, our third installment, we expand on these ideas, examining what happens when independent lines of accelerating technology (artificial intelligence, for example) converge with other independent lines of accelerating technology (augmented reality, for example). Sure, AI is powerful. Augmented reality is too. But it's their convergence that is reinventing retail, advertising, entertainment, and education—just to name a few of the major transformations still ahead.

As we'll see in the pages to come, these convergences are happening at an ever-increasing rate. This has turbo-boosted both the rate of change in the world and the scale of that change. So buckle up is our point, because you're in for a wild ride.

The inspiration for this book emerged from the authors' firsthand experience of this ride, a palpable acceleration in the pace of change in their own businesses and the world. Diamandis is working on his twenty-second startup, the most recent of them in the fields of longevity and healthcare. Coupled with his leadership roles in Singularity University, XPRIZE, Bold Capital Partners, and Abundance 360, this frenzied daily dance provides him a continuous infusion of converging technological insights.

Steven has encountered this acceleration in both his work as an author, where this book marks his sixth devoted to the topic of technology, and as the founder and executive director of the Flow Research Collective, where he focuses on the research and training of peak performance—that is, the very psychological tools we humans need to thrive in this world of amplified change.

As authors, we'd also like to say that this wild ride has been something of a challenge. In the pages ahead, you'll find descriptions of cutting-edge researchers and companies built atop their research. However, keeping pace has not been easy. Companies on the cutting edge when we started writing in early 2018, were often edged out by other companies by the time we completed the writing in late 2019. In other words, while the names are important, those names might change. The heart of this book belongs to the overarching trends of convergence and the transformative impact they're having on business, industry, and our lives.

There is little doubt that the decade to come will be filled with radical breakthroughs and world-changing surprises. As the chapters ahead make clear, every major industry on our planet is about to be completely reimagined. For entrepreneurs, for innovators, for leaders, for anyone sufficiently nimble and adventurous, the opportunities will be incredible. It will be both a future that's faster than you think and arguably the greatest display of imagination rendered visible the world has yet seen. Welcome to an era of extraordinary.

PART ONE

THE POWER OF CONVERGENCE

CHAPTER ONE

Convergence

Flying Cars

The Skirball Cultural Center sits just off the 405 Freeway, on the northern edge of Los Angeles. Built atop the thin spine of the Santa Monica Mountains, the Center offers spectacular views in nearly every direction, except for the freeway below—which is bumper-to-bumper for miles on end.

Of course it is.

In 2018, for the sixth straight year, Los Angeles earned the dubious honor of being the most gridlocked metropolis in the world, where the average driver spends two-and-a-half working weeks a year trapped in traffic. Yet help may be on its way. In May 2018, the Skirball Center was ground zero for Uber Elevate, the ridesharing company's radical plan for solving this traffic: their second annual flying car conference.

Inside the Skirball, giant screens displayed a night sky dotted with stars that slowly faded into a blue sky dotted with clouds. Beneath the clouds, it was standing room only. The event had attracted a motley crew of the power elite: CEOs, entrepreneurs, architects, designers, technologists, venture capitalists, government officials, and real estate magnates. Nearly a thousand in total, dressed in everything from Wall

Street slick to eternally casual Friday, all gathered to witness the birth of a new industry.

To kick off the conference, Jeff Holden, Uber's (now former) chief product officer, took the stage. With curly brown hair and a gray Uber Air polo shirt, Holden had a boyish demeanor that belied his actual role in the affair. This event, in fact, the entire concept of getting Uber off the ground, was Holden's vision.

It was quite a vision.

"We've come to accept extreme congestion as part of our lives," said Holden.[1] "In the U.S., we have the honor of being home to ten of the world's twenty-five most congested cities, costing us approximately $300 billion in lost income and productivity. Uber's mission is to solve urban mobility. . . . Our goal is to introduce an entirely new form of transportation to the world, namely urban aviation, or what I prefer to call 'aerial ridesharing.' "

Aerial ridesharing might sound like sci-fi cliché, but Holden had a solid track record of disruptive innovation. In the late 1990s, he followed Jeff Bezos from New York to Seattle to become one of the earliest employees at Amazon. There, he was put in charge of implementing the then zany idea of free two-day shipping for a flat annual membership fee. It was an innovation that many thought would bankrupt the company. Instead, Amazon Prime was born, and today, 100 million Prime members later, that zany idea accounts for a significant portion of the company's bottom line.

Next, Holden went to another startup, Groupon—which is hard to remember as a disruptive enterprise today, but was then part of the first wave of "power to the people" internet companies. From there, he went to Uber, where, despite the turmoil the company experienced, Holden strung together a series of unlikely wins: UberPool, Uber Eats, and, most recently, Uber's self-driving car program. So when he proposed an

---

1 Unless otherwise noted in either the text or in the endnotes, all quotes come from direct interviews with sources or, as in this case, the author(s) attended the event in question.

even zanier product line—that Uber take to the skies—it wasn't all that surprising that the company's leadership took him seriously.

And for good reason. The theme of the second annual Uber Elevate wasn't actually flying cars. The cars have already arrived. Instead, the theme of the second Uber Elevate was the path to scale. And the more critical point: That path is a lot shorter than many suspect.

By mid-2019, over $1 billion had been invested in at least twenty-five different flying car companies. A dozen vehicles are currently being test-flown, while another dozen are at stages ranging from Power-Point to prototype. They come in all shapes and sizes, from motorcycles stacked atop oversized fans, to quadcopter drones scaled up to human size, to miniature space-pod airplanes. Larry Page, cofounder and CEO of Alphabet, Google's parent company, was among the first to recognize their potential, personally funding three companies, Zee Aero, Opener and Kitty Hawk. Established players like Boeing, Airbus, Embraer, and Bell Helicopter (now just called Bell, a reference to the future disappearance of the helicopter itself) are also in the game. Thus, for the first time in history, we're past the point of talking about the possibility of flying cars.

The cars are here.

"Uber's goal," explained Holden from the stage, "is to demonstrate flying car capability in 2020 and have aerial ridesharing fully operational in Dallas and LA by 2023." But then Holden went even further: "Ultimately, we want to make it economically irrational to own and use a car."

How irrational? Let's look at the numbers.

Today, the marginal cost of car ownership—that is, not the purchase price, but everything else that goes with a car (gas, repairs, insurance, parking, etc.)—is 59 cents per passenger mile. For comparison, a helicopter, which has many more problems than just cost, covers a mile for about $8.93. For its 2020 launch, according to Holden, Uber Air wants to reduce that per mile price to $5.73, then rapidly drive it down to $1.84. But Uber's long-term target is the game-changer—44 cents per mile—or cheaper than the cost of driving.

And you get a lot per mile. Uber's main interest is in "electric vertical take-off and landing vehicles"—or eVTOLs for short. eVTOLs are being developed by a plethora of companies, but Uber has very particular needs. For an eVTOL to qualify for their aerial ridesharing program, it must be able to carry one pilot and four passengers at a speed of over 150 mph for three continuous hours of operation. While Uber envisions twenty-five miles as its shortest flight (think Malibu to downtown Los Angeles), these requirements allow you to leap from northern San Diego to southern San Francisco in a single bound. Uber already has five partners who have committed to delivering eVTOLs that meet these specs, with another five or ten still to come.

But the vehicles alone won't make car ownership irrational. Uber has also partnered with NASA and the FAA to develop an air traffic management system to coordinate their flying fleet. They've also teamed up with architects, designers, and real estate developers to design a string of "mega-skyports" needed for passengers to load and unload and for vehicles to take off and land. Just like with the flying cars, Uber doesn't want to own these skyports, they want to lease them. Once again, they have very specific needs. To qualify as Uber-ready, a mega-skyport must be able to recharge vehicles in seven to fifteen minutes, handle one thousand takeoffs and landings per hour (four thousand passengers), and occupy no more than three acres of land—which is small enough to sit atop old parking garages or on the roofs of skyscrapers.

Put all this together, and by 2027 or so, you'll be able to order up an aerial rideshare as easily as you do an Uber today. And by 2030, urban aviation could be a major mode of getting from A to B.

But all of this raises a fundamental question: Why now? Why, in the late spring of 2018, are flying cars suddenly ready for prime time? What is it about this particular moment in history that has turned one of our oldest science fiction fantasies into our latest reality?

After all, we've been dreaming of *Blade Runner* hover cars and *Back to the Future* DeLorean DMC-12s for millennia. Vehicles capable of flight date back to the "flying chariots" in the Ramayana, an eleventh-century Hindu text. Even the more modern incarnations—that is, ones built around the internal combustion engine—have been around for a while. The 1917 Curtiss Autoplane, the 1937 Arrowbile, the 1946 Airphibian, the list goes on. There are over a hundred different patents on file in the US for "roadable aircraft." A handful have flown. Most have not. None have delivered on the promise of *The Jetsons*.

In fact, our ire at this lack of delivery has become a meme unto itself. At the turn of the last century, in a now famous IBM commercial, comedian Avery Brooks asked: "It's the year 2000, but where are the flying cars? I was promised flying cars. I don't see any flying cars. Why? Why? Why?" In 2011, in his "What Happened to the Future?" manifesto, investor Peter Thiel echoed this concern, writing: "We wanted flying cars, instead we got 140 characters."

Yet, as should be clear by now, the wait is over. The Flying Cars Are Here. And the infrastructure's coming fast. While we were sipping our lattes and checking our Instagram, science fiction became science fact. And this brings us back to our initial question: Why now?

The answer, in a word: Convergence.

Converging Technology

If you want to understand convergence, it helps to start at the beginning. In our earlier books, *Abundance* and *BOLD*, we introduced the notion of exponentially accelerating technology; that is, any technology that doubles in power while dropping in price on a regular basis. Moore's Law is the classic example. In 1965, Intel founder Gordon Moore noticed that the number of transistors on an integrated circuit had been doubling every eighteen months. This meant every year-and-a-half computers got twice as powerful, yet their cost stayed the same.

Moore thought this was pretty astounding. He predicted this trend might last a few more years, maybe five, possibly ten. Well, it's been twenty, forty, going on sixty years. Moore's Law is the reason the smartphone in your pocket is a thousand times smaller, a thousand times cheaper, and a million times more powerful than a supercomputer from the 1970s.

And it's not slowing down.

Despite reports that we are approaching the heat death of Moore's Law—which we'll address in the next chapter—in 2023 the average thousand-dollar laptop will have the same computing power as a human brain (roughly 10^{16} cycles per second). Twenty-five years after that, that same average laptop will have the power of all the human brains currently on Earth.

More critically, it's not just integrated circuits that are progressing at this rate. In the 1990s, Ray Kurzweil, the director of engineering at Google and Peter's cofounding partner in Singularity University, discovered that once a technology becomes digital—that is, once it can be programmed in the ones and zeroes of computer code—it hops on the back of Moore's Law and begins accelerating exponentially.

In simple terms, we use our new computers to design even faster new computers, and this creates a positive feedback loop that further accelerates our acceleration—what Kurzweil calls the "Law of Accelerating Returns." The technologies now accelerating at this rate include some of the most potent innovations we have yet dreamed up: quantum computers, artificial intelligence, robotics, nanotechnology, biotechnology, material science, networks, sensors, 3-D printing, augmented reality, virtual reality, blockchain, and more.

But all of this progress, however radical it may seem, is actually old news. The new news is that formerly independent waves of exponentially accelerating technology are beginning to converge with other independent waves of exponentially accelerating technology. For example, the speed of drug development is accelerating, not only because biotechnology is progressing at an exponential rate, but because artificial intelligence, quantum computing, and a couple other exponentials are converging on the field. In other words, these waves are starting to overlap, stacking atop one another, producing tsunami-sized behemoths that threaten to wash away most everything in their path.

When a new innovation creates a new market and washes away an existing one, we use the term "disruptive innovation" to describe it. When silicon chips replaced vacuum tubes at the beginning of the digital age, this was a disruptive innovation. Yet, as exponential technologies converge, their potential for disruption increases in scale. Solitary exponentials disrupt products, services, and markets—like when Netflix ate Blockbuster for lunch—while convergent exponentials wash away products, services, and markets, as well as the structures that support them.

But we're getting ahead of ourselves. The rest of this book is devoted to these forces and their rapid and revolutionary impact. Before we dive deeper into that tale, let's first examine convergence through a more manageable lens, returning to our initial question about flying cars: Why now?

To answer that, let's examine the three basic requirements any Uber eVTOL will have to meet: safety, noise, and price. Helicopters, which are the closest model anyone has for a flying car, have been around for nearly eighty years—Igor Sikorsky built the world's first one in 1939—yet they can't come close to satisfying these requirements. Besides being loud and expensive, they have that bad habit of falling out of the sky. So why are Bell, Uber, Airbus, Boeing, and Embraer—just to name a few—bringing aerial taxis to market today?

Once again: Convergence.

Helicopters are loud and dangerous because they use a single gargantuan rotor to generate lift. Unfortunately, the tip-speed of that single rotor produces exactly the right *thud-thud-thud* frequency to annoy pretty much anyone with ears. And they're dangerous because, if that rotor fails, well, gravity plays for keeps.

Now imagine, instead of one main rotor overhead, a bunch of smaller rotors—like a row of small fans beneath a plane's wing—whose combination generates enough lift to fly, but pumps out a lot less noise. Better yet, imagine if this multi-rotor system could fail gracefully, landing safely even if a couple rotors stopped working at once. Add to this design a single wing that enables speeds of 150 mph or more. All great ideas, except, thanks to their terrible power-to-weight ratios, gasoline-powered engines make none of this possible.

Enter distributed electric propulsion, or DEP for short.

Over the past decade, a surge in demand for commercial and military drones has pushed roboticists (and drones are just flying robots) to envision a new kind of electromagnetic motor: extremely light, stealthily quiet, and capable of carrying heavy loads. To design that motor, engineers relied on a trilogy of converging techs: first, machine learning advances that allowed them to run enormously complicated flight simulations, then materials science breakthroughs that let them create parts both light enough for flying and durable enough for safety, and last, new manufacturing techniques—3-D printing—that can create these motors and rotors at any scale. And talk about functionality: These electric engines are 95 percent efficient compared to gasoline's 28 percent.

But flying a DEP system is another story. Adjusting a dozen motors in microsecond intervals is beyond a human pilot's skill. DEP systems are "fly-by-wire"—that is, computer controlled. And what produces that level of control? Another swarm of converging technologies.

First, an AI revolution gave us the computational processing horsepower to take in an ungodly amount of data, make sense of it in microseconds, and manipulate a multitude of electric motors and aircraft control surfaces accordingly, in real time. Second, to sweep in all that data, you need to replace the pilot's eyes and ears with sensors capable of processing gigabits of information at once. That means GPS, LIDAR, radar, an advanced visual imaging suite, and a plethora of microscopic accelerometers—many of which are the dividends of a decade of smartphone wars.

Finally, you'll need batteries. They'll have to last long enough to overcome range anxiety—or the fear of running out of juice while running errands—and generate enough oomph, or what engineers call "power density," to lift the vehicle, a pilot, and four passengers off the ground. To achieve this lift, the minimum requirement is 350 kilowatt hours per kilogram. This was out of reach, until recently. Thanks to the explosive growth of both solar power and electric cars, there's now a bigger need for better energy storage systems, resulting in a next generation of lithium-ion batteries with increased range, and, as an added bonus, enough power to lift flying cars.

In the aerial ridesharing equation, we've solved safety and noise, but price still requires a few more innovations. There's also the not small issue of manufacturing enough eVTOLs for Uber's program. To be able to meet Uber's outsized demand at an affordable price would require suppliers to produce aircraft faster than during WWII, when a still unbroken record of eighteen thousand B24 fighters were pumped out over two years—or, at its peak, one plane every sixty-three minutes.

For this to happen—which is what it would take to make flying cars a mainstream reality, not an elitist luxury—we need another trio of convergences. To start, computer-aided design and simulation need to become deft enough to draft the airfoils, wings, and fuselages required for commercial flight. At the same time, material science has to produce carbon fiber composites and complex metal alloys that are light enough for flight yet durable enough for safety. Finally, 3-D printers have to become fast enough to turn these new materials into usable parts so that all previous aircraft manufacturing records are shattered. In other words, *exactly* where we are today.

Sure, you can play this game with any new technology. Socks couldn't be invented until a materials revolution turned plant fibers into soft fabrics and a tool-making revolution turned animal bones into sewing needles. This is progress, of course, but linear in nature. It took thousands of years to get from these first steps in sock-dom to the next major innovation: the domestication of animals (which gave us sheep's wool). And thousands more years for electricity to bring sock-making to scale.

But the blurry acceleration we're witnessing today—that is, the answer to "Why now?"—is the result of a dozen different technologies converging. It's progress at a rate that we've not seen before. And this is a problem for us.

The human brain evolved in an environment that was local and linear. Local, meaning most everything that we interacted with was less than a day's walk away. Linear, meaning the rate of change was exceptionally slow. Your great-great-great-grandfather's life was roughly the same as his great-great-grandson's life. But now we live in a world that is global and exponential. Global, meaning if it happens on the other side of the planet, we hear about it seconds later (and our computers hear about it only milliseconds later). Exponential, meanwhile, refers to today's blitzkrieg speed of development. Forget about the difference between generations, currently mere months can bring a revolution. Yet our brain—which hasn't really had a hardware update in two hundred thousand years—wasn't designed for this scale or speed.

And if we struggle to track the growth of singular innovations, we're downright helpless in the face of converging ones. Put it this way, in "The Law of Accelerating Returns," Ray Kurzweil did the math and found that we're going to experience twenty thousand years of technological change over the next one hundred years. Essentially, we're going from the birth of agriculture to the birth of the internet twice in the next century. This means paradigm-shifting, game-changing, nothing-is-ever-the-same-again breakthroughs—such as affordable aerial ridesharing—will not be an occasional affair. They'll be happening all the time.

It means, of course, that flying cars are just the beginning.

More Transportation Options

Autonomous Cars

A little over a century ago, another transportation transformation was under way. The triple threat convergence of the internal combustion engine, the moving assembly line and the emerging petroleum industry was together driving—pardon the pun—the horse-and-buggy business out of business.

The first bespoke cars hit the roads around the tail end of the nineteenth century, but Ford's 1908 introduction of the mass-produced

Model T marked the real tipping point. Just four years later, New York traffic surveys counted more cars than horses on the road. And while the speed of this shift was breathtaking, in retrospect it wasn't unexpected. Whenever a new technology offers a tenfold increase in value—cheaper, faster and better—there's little that can slow it down.

In the decades that followed Ford's invention, with a Cambrian explosion of accouterments, the car reshaped our world: stoplights and stop signs, interstate highways and multilevel interchanges, parking lots and parking garages, gas stations on every corner, the drive-thru, car washes, suburbs, smog and gridlock. But even as we witness the birth of aerial ridesharing—which seems likely to replace multiple parts in this system—a different revolution threatens it entirely: autonomous cars.

While the first driverless car was a radio-controlled "American wonder" that navigated the streets of New York City back in the 1920s, this was little more than an oversized toy. Its more modern incarnation emerged from the military's desire for a risk-free way to resupply troops. Roboticists began trying to meet this need in the 1980s; car companies started paying attention in the nineties. Many date the pivotal breakthrough to 2004, when the Defense Advanced Research Projects Association (DARPA) created a driverless car competition—the DARPA Grand Challenge—to turbocharge development.

The competition did its job. A decade later, most major car companies and more than a few major tech companies had autonomous car programs up and running. By the middle of 2019, dozens of vehicles had logged millions of miles on California roads. Traditional automotive players like BMW, Mercedes and Toyota were competing for this emerging market with tech giants like Apple, Google (via Waymo), Uber, and Tesla, trying out different designs, gathering data, and honing neural networks.

Out of these, Waymo seems well positioned for early market dominance. Formerly Google's self-driving car project, Waymo began its work in 2009 by hiring Sebastian Thrun, the Stanford professor who won the DARPA Grand Challenge. Thrun helped develop the AI system that would become the brains behind Waymo's self-driving fleet. About ten years later, in March 2018, Waymo purchased that fleet, buying twenty thousand sporty, self-driving Jaguars for its forthcoming ride-hailing service. With this many cars, Waymo intends to deliver a million trips *per day* in 2020 (this might be ambitious but Uber currently delivers 15 million rides a day). To understand the importance of this figure or anything close to it, consider that the more miles an autonomous car drives, the more data it gathers—and data is the gasoline of the driverless world.

Since 2009, Waymo's vehicles have logged over 10 million miles. By 2020, with twenty thousand Jaguars doing hundreds of thousands of daily trips, they'll be adding an extra million miles or so every day. All of those miles matter. As autonomous vehicles drive, they gather information: positions of traffic signs, road conditions, and the like. More information equals smarter algorithms equals safer cars—and this combination is the very edge needed for market domination.

To compete with Waymo, General Motors is making up for lost time with big dollars. In 2018, it poured $1.1 billion into GM Cruise, its self-driving division. A few months later, it took an additional $2.25 billion investment from the Japanese conglomerate Softbank, just months after Softbank had taken a 15 percent position in Uber. With all of this capital flying around, with all these heavy hitters involved, how fast will this transformation occur?

"Faster than anyone expects," says Jeff Holden (who also founded Uber's AI lab and autonomous car group). "Already, over 10 percent of millennials have opted for ridesharing over car ownership, but this is just the beginning. Autonomous cars will be four to five times cheaper—they make owning a car not only unnecessary, but also expensive. My guess, within ten years, you'll probably need a special permit to drive a human-operated car."

For consumers, the benefits of this transformation are many. Most Americans will tolerate a commute of thirty minutes or less, but with a robo-chauffeur behind the wheel and a car that can become anything—a bedroom, a meeting room, a movie theater—you might not mind living farther afield, where lower-cost real estate lets you buy more house for less money. Giving up that car allows you to turn your garage into a spare bedroom, your driveway into a rose garden, and you won't need to buy gas again—ever. The cars are electric, and they recharge themselves at night. No more hunting for parking spots, or fretting over parking tickets. No speeding tickets either. Or drunk driving. NOTE: City revenues could plunge.

All of these trends are disruptive in nature. But they pale in comparison to two larger forces for change: first, demonetization, or the removal of cash from the equation. Ridesharing autonomous cars price out at 80 percent cheaper than individual car ownership, and they come equipped with a robo-chauffeur. Second: saved time. The average U.S. roundtrip commute is 50.8 minutes of hair-pulling, mind-numbing drudgery that can be repurposed for sleep, reading, tweeting, sex . . . whatever your pleasure.

For big car manufacturers, these developments spell the beginning of the end, especially for those selling car-as-possession rather than car-as-service. In 2019, there were a hundred plus automotive brands in existence. Over the next ten years, we can expect auto industry consolidation as exponential technology takes direct aim at Detroit, Germany, and Japan.

Car usage rates will be the first driver of this consolidation. Today, the average car owner drives their vehicle less than 5 percent of the time, and a family of two adults typically has two cars. Thus, a single autonomous car can serve a half-dozen families a day. However you work those numbers, this dramatic increase in cooperative efficiency will significantly reduce the need for new car production.

Functionality is the second driver. In a ridesharer's marketplace, the companies that collect the most data and assemble the biggest fleets are the ones that will offer the lowest wait times and cheapest rides. Cheap and quick are the two biggest factors impacting consumer choice in this kind of market. What brand of car ridersharers are sharing matters a lot less. Most of the time, if the vehicle is clean and neat, consumers won't even notice what brand the car is—similar to how most of us feel about Uber or Lyft today. So, if a half-a-dozen different vehicles are all it takes to please the customer, then a wave of car company extinction is going to follow our wave of car company consolidation.

Big auto won't be the only industry impacted. America has almost half-a-million parking spaces. In a recent survey, MIT professor of urban planning Eran Ben-Joseph reported that, in many major US cities, "parking lots cover more than a third of the land area," while the nation as a whole has set aside an area larger than Delaware and Rhode Island combined for our vehicles. But if car-as-service replaces car-as-thing-you-have-to-park, then we're going to be looking at a huge commercial real estate boom as all those lots get repurposed. Then again, a lot of them could become skyports. Whatever the case, transportation ten years from today is going to look radically different—and this prediction doesn't include everything that happened after Elon Musk lost his temper.

Hyperloop

On an empty swatch of desert outside of Las Vegas, perched atop a high-tech stretch of track, a sleek silver pod begins to quiver. Less than a second later, it's not just moving, it's a hundred-mile-per-hour blur. Ten seconds after, it's zipping down the Virgin Hyperloop One Development Track at 240 mph. If these tracks continued—as they someday will—this high-speed train would take you from Los Angeles to San Francisco in the time it takes to watch a sitcom.

Hyperloop is the brainchild of Elon Musk, just one in a series of transportation innovations from a man determined to leave his mark on the industry. In *BOLD*, we explored his first two forays: SpaceX, his rocket company, and Tesla, his electric car company. SpaceX helped revitalize aerospace commercial launches, turning a fantasy into a billion-dollar industry. Tesla's rapid rise to prominence, meanwhile, shook the major automotive companies out of their electric car apathy. As a result, all have begun phasing out gas guzzlers in favor of fully rechargeable fleets.

And both of these companies began to flourish before Musk got irritated.

In 2013, in an attempt to shorten the long commute between Los Angeles and San Francisco, the California state legislature proposed a $68 billion budget allocation on what appeared to be the slowest and most expensive bullet train in history. Musk was outraged. The cost was too high, the train too sluggish. Teaming up with a group of engineers from Tesla and SpaceX, he published a fifty-eight-page concept paper for "The Hyperloop," a high-speed transportation network that used magnetic levitation to propel passenger pods down vacuum tubes at speeds up to 760 mph. If successful, it would zip you across California in thirty-five minutes—or faster than commercial jets.

Musk's idea wasn't entirely new. Sci-fi dreamers have long envisioned high-speed travel through low-pressure tubes. In 1909, rocketry pioneer Robert Goddard proposed a vacuum train concept similar to the Hyperloop. In 1972, the RAND Corporation extended this into a supersonic underground railway. But just like flying cars, turning sci-fi into sci-fact required a series of convergences.

The first of these convergences wasn't technological. Rather, it was about the people involved. In January 2013, Musk and venture capitalist Shervin Pishevar were on a humanitarian mission to Cuba when they fell into a discussion about the Hyperloop. Pishevar saw possibilities, Musk saw overwhelm. He was irate enough to publish a white paper, but way too busy to start another company. So Pishevar, with Musk's blessing, decided to do so himself. With Peter (one of your authors), former White House Deputy chief of staff for Obama Jim Messina, and tech entrepreneurs Joe Lonsdale and David Sacks as founding board members, Pishevar created Hyperloop One. A couple of years after that, the Virgin Group invested in the idea, Richard Branson was elected chairman, and Virgin Hyperloop One was born.

The other required convergences were technological in nature. "The Hyperloop exists," says Josh Giegel, the cofounder and chief technology officer for Hyperloop One, "because of the rapid acceleration of power electronics, computational modeling, material sciences, and 3-D printing. Computational power has increased so much that we can now run hyperloop simulations on the cloud, testing the whole system for safety and reliability. And manufacturing breakthroughs ranging from the 3-D printing of electromagnetic systems to the 3-D printing of large concrete structures have changed the game in terms of price and speed."

These convergences are why, in various stages of development, there are now ten major Hyperloop One projects spread across the globe. Chicago to DC in thirty-five minutes. Pune to Mumbai in twenty-five minutes. According to Giegel: "Hyperloop is targeting certification in 2023. By 2025, the company plans to have multiple projects under construction and running initial passenger testing."

So think about this timetable: Autonomous car rollouts by 2020. Hyperloop certification and aerial ridesharing by 2023. By 2025— going on vacation might have a totally different meaning. Going to work most definitely will. And Musk was just getting started.

The Boring Company

Elon Musk's main residence in Los Angeles is located in Bel Air, a seventeen-mile trek from SpaceX's Hawthorne-based offices. On the best of days, his commute takes thirty-five minutes—but December 17, 2016 (coincidentally the anniversary of the first Wright brothers flight), was not the best of days. The 405 was at a dead stop, and the pile-up pushed Musk over the edge. It also gave him time to tweet:

> @elonmusk—17 Dec 2016: "Traffic is driving me nuts. Am going to build a tunnel boring machine and just start digging . . ."
>
> @elonmusk—17 Dec 2016: "It shall be called 'The Boring Company' "
>
> @elonmusk—17 Dec 2016: "Boring, it's what we do"
>
> @elonmusk—17 Dec 2016: "I am actually going to do this"

And he did.

Eight months later, on July 20, the anniversary of the Apollo moon landing, Musk tweeted again: "Just received verbal govt approval for The Boring Company to build an underground NY-Phil-Balt-DC Hyperloop. NY-DC in 29 mins." In the spring of 2018, with $113 million of Musk's own money, the Boring Company began boring. They started construction on both ends of the line in DC and New York, while also starting on a 10.3-mile Maryland stretch that will eventually connect the two. And while the tunnel is being designed as "Hyperloop compatible"—meaning it is able to house a Hyperloop—the current plan calls for an interim high-speed train step, where the first trains through will travel around 150 mph (much less than Musk's proposed 700+ mph speeds).

They've also gotten a contract for building a three-stop subway beneath Las Vegas's sprawling convention center—which they hope to have open for the 2021 Consumer Electronics Show. While not a Hyperloop—the distance is just way too short to bother—it does mark the Boring Company's first paying customer.

Finally, while the company has started drilling with conventional machines, Musk has borrowed a page from Tesla's playbook and is now designing electric boring machines that are three times as powerful as the traditional version.

It's also worth noting that all of the innovations discussed in this chapter will work in concert. In the minutes before a Hyperloop pod arrives at a Boring Company–drilled station, the AI behind Uber's aerial ridesharing service and the AI behind Waymo's driverless ridesharing fleet will dispatch a swarm of vehicles to that station in order to take passengers on the next leg of their trip. And if that's not fast enough for you, sometime soon there might just be another option available.

Rockets: LA to Sydney in Thirty Minutes

As if autonomous cars, flying cars, and high-speed trains weren't enough, in September of 2017, speaking at the International Astronautical Congress in Adelaide, Australia, Musk promised that for the price of an economy airline ticket, his rockets will fly you "anywhere on Earth in under an hour."

Musk delivered this promise at the end of an hour-long keynote to five thousand aerospace executives and government officials. The presentation was primarily an update about SpaceX's megarocket, Starship, which was designed to take humans to Mars. The fact that Musk now wanted to use his interplanetary starship for terrestrial passenger delivery was the transportation industry equivalent of Steve Jobs's famous line that (almost) ended his demos: "Wait, wait . . . There's one more thing."

The Starship travels at 17,500 mph. It's an order of magnitude faster than the Concorde. Think about what this actually means: New York to Shanghai in thirty-nine minutes. London to Dubai in twenty-nine minutes. Hong Kong to Singapore in twenty-two minutes. What's not to like?

So how real is the Starship?

"We could probably demonstrate this [technology] in three years," Musk explained, "but it's going to take a while to get the safety right. It's a high bar. Aviation is incredibly safe. You're safer on an airplane than you are at home."

That demonstration is proceeding as planned. In September 2017, Musk announced his intentions to retire his current rocket fleet, both the Falcon 9 and Falcon Heavy, and replace them with the Starships in the 2020s. Less than a year later, LA mayor Eric Garcetti tweeted that SpaceX was planning to break ground on an eighteen-acre rocket production facility near the port of LA. And April of 2019 marked a bigger milestone: the very first test flights of the rocket. Thus, sometime in the next decade or so, "off to Europe for lunch" may become a standard part of our lexicon.

Seeing into the Future

It's about to get personal. Before the end of the next decade, this transportation revolution will impact some of the most intimate aspects of our lives. Where we choose to live and work, how much free time we have, how we spend that time. It will change how cities look and feel, the size of the "local" dating pool, the demographics of the "local" school district— the list goes on and on.

Yet, try to visualize that "on and on." Seriously. Put down this book, close your eyes, and ask yourself a question: How would this transportation transformation change your life? Start small. Consider your day. What errands will you run? What stores will you visit?

Are you sure about that?

This last question may seem innocuous, but think about it this way: In 2006, retail was booming. Sears was worth $14.3 billion, Target $38.2 billion, and Walmart a whopping $158 billion. Meanwhile, an upstart retailer named Amazon was at $17.5 billion. Now fast-forward a decade. What's changed?

Hard times hit Main Street. By 2017, Sears had lost 94 percent of its value, ending the decade worth $0.9 billion, before promptly going out of business. Target did better, finishing up at $55 billion. Walmart did the best, going up to $243.9 billion. But Amazon? The Everything Store closed out the era worth $700 billion (today $800 billion). And it's a fairly safe bet that your life changed as a result.

But all Amazon did to change your life was use a new technology, the internet, to expand upon an old technology, mail-order catalogs. The transportation transformation headed our way sits at the convergence of a half-dozen exponential technologies and the confluence of a half-dozen markets. Not easy to picture all that overlapping impact, is it?

It's not easy for any of us. Studies done with fMRI show that when we project ourselves into the future something peculiar happens: The medial prefrontal cortex shuts down. This is a part of the brain that activates when we think about ourselves. When we think about other people, the inverse happens: It deactivates. And when we think about absolute strangers, it deactivates even more.

You'd expect that thinking about our future selves would excite the medial prefrontal cortex. Yet the opposite happens. It starts to shut down, meaning the brain treats the person we're going to become as a stranger. And the farther you project into the future, the more of a stranger you become. If, a few paragraphs back, you took the time to think about how the transportation revolution would impact future you, the you that you were thinking of was literally not you.

This is why people have a tough time saving for retirement or staying on a diet or getting regular prostate exams—the brain believes that the person who would benefit from those difficult choices isn't the same one making those choices. This is also why, if you've been reading this chapter and having trouble processing the speed of the change ahead, perhaps fluctuating between "total BS" and "holy crap," well, you're not alone. Couple this with the limitations imposed by our local and linear brains in a global and exponential world, and accurate prediction becomes a considerable problem. Even under normal conditions, these built-in features of our neurobiology make us blind to what's around the bend.

But conditions are not close to "normal." Not only are a dozen exponential technologies beginning to converge, their impact is unleashing a series of secondary forces. These forces range from our increasing access to information, money, and tools, to our considerable uptick in productive time and life expectancy. These forces are another tsunami of change, accelerating our acceleration, amping up the speed and scale of the coming disruption.

Which is both good and bad news.

The bad has less to do with what's coming and more to do with our (in)ability to adapt to change. A slew of studies have shown that the convergence of AI and robotics could threaten a significant percentage of America's workforce over the next few decades. That's tens of millions of people who will have to be retrained and retooled if we hope to keep pace. The good news is what's on the other side of that retraining.

Every time a technology goes exponential, we find an internet-sized opportunity tucked inside. Think about the internet itself. While it seemingly decimated industries—music, media, retail, travel, and taxis—a study by McKinsey Global Research found the net created 2.6 new jobs for each one it extinguished.

Over the next decade, we'll see these kinds of opportunities arise in dozens of industries. As a result, if the internet is our benchmark, more wealth could be created over the next ten years than was over the previous century. Entrepreneurs—including, thankfully, environmentally and socially conscious entrepreneurs—have never had it so good. The time it takes to raise seed capital has shrunk from years to minutes. Unicorn formation, or the time it takes to go from "I've got a neat idea" to "I run a billion-dollar company," was once a two-decade long shot. Today, in some cases, it's nothing more than a one-year adventure.

Unfortunately, established organizations will have a hard time keeping pace. Our biggest companies and government agencies were designed in another century, for purposes of safety and stability. Built to last, as the saying goes. They were not built to withstand rapid, radical change. This is why, according to Yale's Richard Foster, 40 percent of today's Fortune 500 companies will be gone in ten years, replaced, for the most part, by upstarts we've not yet heard of.

Institutions are similarly suffering. The educational system was an eighteenth-century invention, designed to batch-process children and prepare them for a life working in factories. That's not today's world, which explains why this system is failing to meet our current needs—and it's not the only institution under duress.

Why are divorce rates so high? One reason is that marriage was created over four thousand years ago, when we got hitched as teens and death came by forty. The institution was designed for a twenty-year maximum commitment. But thanks to advances in healthcare and lifespan, we're now looking at a half century of togetherness—which puts a whole new spin on " 'til death do us part."

The point is this: Being able to see around the corner of tomorrow and being agile enough to adapt to what's coming have never been more important. And, in three parts, that's exactly what this book will do.

In Part One, we'll explore ten technologies currently on exponential growth curves, examining where they are today and where they're going. We'll also assess a series of secondary forces—call them technological shock waves—and see how they're further accelerating the rate of change in the world and amplifying the scale of its impact.

In Part Two, focusing on nine industries, we'll see how converging technologies are reshaping our world. From the future of education and entertainment to the transformation of healthcare and business, this portion provides a blueprint for tomorrow, a map of the major shifts coming to society, and a playbook for anyone interested in surfing that wave.

In Part Three, we move to the bigger picture, looking at a series of environmental, economic, and existential risks that threaten the progress we're about to make. Next, we'll expand our view from the decade ahead to the full century, focusing on five great migrations—economic relocations, climate-change upheavals, virtual worlds explorations, outer space colonization, and hive-mind collaborations—that will play now-you-see-it-now-you-don't with, well, just about everything.

But before we do all that, as Steve Jobs liked to say: Wait, wait . . . There's one more thing.

Avatars

It's 2028 and you're having breakfast at home in Cleveland, Ohio. You stand up, kiss the kids goodbye, and head out the door. Today, it's a meeting in downtown New York. Your personal AI knows your schedule so has an Uber autonomous on standby. As you walk outside, the self-driving car pulls into your driveway.

Time elapsed? Less than ten seconds.

Because you're wearing a sleep sensor—and your AI also knows you didn't get much rest last night—it's the perfect opportunity for a catnap. And your Uber provides nothing less, equipped with a lay-down-flat backseat and a fresh set of sheets.

The car/bed takes you to the local Hyperloop station, where your freshly rested self is transferred into a high-speed pod, then zipped downtown. From the roof of a Cleveland skyscraper, Uber Elevate flies you to one of Manhattan's mega-skyports. You take the elevator down to the ground floor, where another Uber autonomous awaits to take you to your meeting on Wall Street. Total elapsed time, door-to-door: fifty-nine minutes.

To borrow a term from computation, this is a future of "packet-switched humans," where you choose your priority—speed, comfort, or cost—specify your start and end point, and the system does the rest. No fuss, no missed details, and backup options always available.

Wait, wait, there's one more thing.

While the technologies we've discussed will decimate the traditional transportation industry, there's something on the horizon that will disrupt travel itself. What if, to get from A to B, you didn't have to move your body? What if you could quote Captain Kirk and just say: "Beam me up, Scotty."

Well, shy of the *Star Trek* transporter, there's the world of avatars.

An avatar is a second self, typically in one of two forms. The digital version has been around for a couple of decades. It emerged from the video game industry and was popularized by virtual world sites like Second Life and books-turned-blockbusters like *Ready Player One*. A VR headset teleports your eyes and ears to another location, while a set of haptic sensors shifts your sense of touch. Suddenly, you're inside an avatar inside a virtual world. As you move in the real world, your avatar moves in the virtual. Use this technology to give a lecture and you can do it from the comfort of your living room, skipping the trip to the airport, the cross-country flight, and the ride to the conference center.

Robots are the second form of avatars. Imagine a humanoid robot that you can occupy at will. Maybe, in a city far from home, you've rented the bot by the minute—via a different kind of ridesharing company—or maybe you have spare robot avatars located around the country. Either way, put on VR goggles and a haptic suit and you can teleport your senses into that robot. This allows you to walk around, shake hands, and take action—all without having to leave your home.

And like the rest of the tech we've been talking about, even this future isn't very far away. In 2018, All Nippon Airways (ANA) funded the $10 million ANA Avatar XPRIZE to speed the development of robotic avatars. Why? Because ANA knows this is one of the technologies likely to disrupt the airline industry—their industry—and they want to be ready.

To put this in different terms, individual car ownership enjoyed over a century of ascendency. The first real threat it faced, today's ridesharing model, only showed up in the last decade. But that ridesharing model won't even get ten years to dominate. Already, it's on the brink of autonomous car displacement, which is on the brink of flying car disruption, which is on the brink of Hyperloop and rockets-to-anywhere decimation. Plus, avatars. The most important part: All of this change will happen over the next ten years.

Welcome to the future that's faster than you think.

CHAPTER TWO

The Jump to Lightspeed:

Exponential Technologies Part I

Quantum Computing

The coldest place in the universe is located in sunny California. On the outskirts of Berkeley, inside an oversized warehouse, hangs a large white pipe. It's a human-made contraption, a next generation cryogenic refrigerator cooled to .003 Kelvin, or just north of absolute zero.

Back in 1995, astronomers in Chile detected temperatures of 1.15 Kelvin inside the Boomerang Nebula. It was a record-setting discovery: the coldest naturally occurring spot in the cosmos. The white pipe, meanwhile, is nearly a degree cooler—which makes it both the chilliest corner of our universe and the kind of radical freeze needed to hold a qubit in superposition.

Hold a what in a what?

In classical computing, a "bit" is a tiny chunk of binary information, either a one or a zero. A "qubit" is the newer version of this idea, or a quantum bit. Unlike binary bits, which are an either/or scenario, qubits utilize "superposition," which allows them to be in multiple states at once. Think of the two options of flipping a coin: heads or tails. Now think about a spinning coin—where both states flash by at once. That's superposition, only it requires super-cold temperatures to achieve.

Superposition means power. A lot of power. A classical computer requires thousands of steps to solve a hard problem, but a quantum computer can accomplish that same task in two or three steps total. To put this into perspective, IBM's Deep Blue computer that beat Gary Kasparov at chess examined 200 million moves a second. A quantum machine can bump that up to a trillion or more—and that's the kind of power hidden inside that large white pipe.

The pipe belongs to Rigetti Computing, a six-year-old company at the center of one of the most interesting David v. Goliath sagas in technology. Currently, the main contestants in the race toward "quantum supremacy"—that is, the race to build a quantum computer that can solve a problem unsolvable by classical machines—are tech behemoths such as Google, IBM, and Microsoft, academic standouts like Oxford and Yale, the government of China and the US, and the aforementioned Rigetti.

The company got going in 2013, when a physicist named Chad Rigetti decided that quantum computers were a lot closer to prime time than many suspected, and that he wanted to be the one to push the technology over the finish line. So he left a comfortable job as a quantum researcher at IBM, raised over $119 million in funding, and built the coldest pipe in history. Over fifty patent applications later, Rigetti now manufactures integrated quantum circuits that power quantum computers in the cloud. And he's right, this technology does solve one great problem: the end of Moore's Law.

Over the next two chapters, we're going to examine ten exponential technologies that are all beginning to converge. All are surfing Moore's Law, a six-decade wave of rising computational capacity. Transistor power—which is how you measure the size of this wave—is often calculated in FLOPS, or floating operations per second. In 1956, our computers were capable of ten thousand FLOPS. In 2015, this had become one *quadrillion* FLOPS. And it's this trillionfold improvement that's been the most important force driving technology forward.

Yet, over the past few years, Moore's Law has been slowing down. The issue is physics. Improvements to integrated circuits have come via shrinking the space between transistors, which allows us to pack more of them onto a chip. In 1971, channel distance—that is, the distance between transistors—was ten thousand nanometers. By 2000, it was roughly one hundred nanometers. Today, we're closing in on five, which is where the trouble starts. At this microscopic scale, electrons begin to jump around, destroying their ability to calculate. This makes it the hard physical limit of transistor growth—the swan song of Moore's Law.

Except—maybe not.

"Moore's Law was not the first, but the fifth paradigm to provide accelerating price-performance," writes Ray Kurzweil in "The Law of Accelerating Returns." "Computing devices have been consistently multiplying in power (per unit of time) from the mechanical calculating devices used in the 1890 U.S. Census, to Turing's relay-based 'Robinson' machine that cracked the Nazi Enigma code, to the CBS vacuum tube computer that predicted the election of Eisenhower, to the transistor-based machines used in the first space launches, to the integrated-circuit-based personal computer which I used to dictate this essay."

Kurzweil's point is that every time an exponential technology reaches the end of its usefulness, another arises to take its place. And so it is with transistors. Right now, there are a half-dozen solutions to the end of Moore's Law. Alternative uses of materials are being explored, such as replacing silicon circuits with carbon nanotubes for faster switching and better heat dissipation. Novel designs are also in the works, including three-dimensional integrated circuits, which geometrically increase the available surface area. There are also specialized chips that have limited functionality, but incredible speed. Apple's recent A12 Bionic, for example, only runs AI applications, but does so at a blistering nine trillion operations a second.

Yet all of these solutions pale in comparison to quantum computing.

In 2002, Geordie Rose, the founder of an early quantum computer company D-Wave, came up with the quantum version of Moore's Law, what's now known as Rose's Law. The idea is similar: The number of qubits in a quantum computer doubles every year. Yet Rose's Law has been described as "Moore's Law on steroids," because qubits in superposition have way more power than binary bits in transistors. Put it this way: A fifty-qubit computer has sixteen petabytes of memory. That's a lot of memory. If it were an iPod, it would hold fifty million songs. But increase that by a mere thirty qubits and you get something else entirely. If all the atoms of the universe were capable of storing one bit of information, an eighty-qubit computer would have more information storage capacity than all the atoms in the universe.

For this same reason, we have no real idea what innovations will arise once quantum computing actually starts to mature. But what we do know is tantalizing. Because chemistry and physics are quantum processes, computing in qubits will usher in what Oxford's Simon Benjamin calls "a golden age of discovery in new materials, new chemicals, and new drugs." It will also amplify artificial intelligence, remake cybersecurity, and allow us to simulate incredibly complex systems.

How will quantum computing help lead to new drugs, for example?

Chad Rigetti explains:"[The technology] change[s] the economics of research and development. Say you're trying to create a new cancer drug. Instead of building a large-scale wet lab to explore the properties of hundreds of thousands of compounds in test tubes, you're going to be able to do much of that exploration inside a computer." In other words, the gap between neat idea and new drug is about to become a whole lot shorter.

And everyone can play. Quantum computing is already available to the masses. Right now, if you go to the website for Rigetti Computing (www.rigetti.com), you can download Forest, their quantum developer's kit. The kit provides a user-friendly interface to the quantum world. With it, almost anyone can write programs that can be run on Rigetti's thirty-two-qubit computer. Over 120 million programs have been run.

The development of a user-friendly interface for quantum computing marks a critical inflection point. Maybe, *the* critical inflection point, but this takes a little explaining. . . .

In *BOLD*, we introduced "the Six Ds of Exponentials," or the growth cycle of exponential technologies: Digitalization, Deception, Disruption, Demonetization, Dematerialization, and Democratization. Each represents a crucial phase of development for an exponential technology, one that always leads to enormous upheaval and opportunity. Since understanding these stages will be indispensable to understanding the evolution of quantum computing (and the other technologies we'll be discussing), they're worth taking a moment to review:

Digitalization: Once a technology becomes digital, meaning once you can translate it into the 1s and 0s of binary code, it jumps on the back of Moore's Law and begins accelerating exponentially. Soon, with quantum, this will be on the back of Rose's Law and an even wilder ride.

Deception: Exponentials typically generate a lot of hype when first introduced. Because early progress is slow (when plotted on a curve, the first few doublings are all below 1.0), these technologies spend a long time failing to live up to the hype. Think about the initial days of Bitcoin. Back then, most people thought crypto was a novelty toy for übergeeks or a way to buy illegal drugs online. Today, it's a reinvention of our financial markets. This is a classic example of the deceptive phase.

Disruption: This is what happens when exponentials really start to impact the world, when they begin disrupting existing products, services, markets, and industries. An example is 3-D printing, a single exponential technology that threatens the entire $10 trillion manufacturing sector.

Demonetization: Where a product or service once had a cost, now money vanishes from the equation. Photographs were once expensive.

You took limited numbers of pictures because film and developing that film cost a lot of money. But once photos became digital, those costs vanished. Now you take photos without thinking of them, and the difficulty comes in sorting through too many options.

Dematerialization: Now you see it, now you don't. This is when the products themselves disappear. Cameras, stereos, video game consoles, TVs, GPS systems, calculators, paper, matchmaking as we knew it, etc. These once independent products are now standard fare on any smartphone. Wikipedia dematerialized the encyclopedia; iTunes dematerialized the music store. Etc.

Democratization: This is when an exponential scales and goes wide. Cell phones were once brick-sized instruments available only to a wealthy few. Today, almost everyone has one, making it nearly impossible to find anywhere in the world untouched by this technology.

So what does this mean for quantum computing? Well, in light of this growth cycle, a user-friendly interface is what bridges the gap between a technology's deceptive and disruptive phase. Consider the internet. In 1993, Marc Andreesen designed Mosaic, the first user-friendly interface for the internet (what became the Netscape browser). Before he did, there were twenty-six websites online. A few years later, there were hundreds of thousands; a few years after that, millions. This is the real power of a user-friendly interface—it democratizes technology. By allowing nonexperts to play, an interface allows a technology to scale. And fast. Thus the fact that 1.5 million programs have been run on Rigetti's Forest—its user-friendly interface to the quantum world—tells us that radical change is right around the corner.

Artificial Intelligence

In 2014, Microsoft released a chatbot in China. Her name was Xiaoice (pronounced Shao-ice), and her mission was something of a test. Unlike most personal AIs, which tend to be designed for task completion, Xiaoice was optimized for friendliness. Instead of getting the job done fast, her goal was to keep that conversation going. And since Xiaoice was designed to respond like a seventeen-year-old girl, she isn't always polite.

Sarcastic, ironic, and often surprising? Yeah, there's plenty of that. For example, while Xiaoice was built with neural nets—a technology we'll explain in a moment—when asked if she understands how neural nets work, she responds, "Yeah, magnets!"

What's more surprising is how much people like talking to Xiaoice. Since her debut, Xiaoice has had over 30 billion conversations with over 100 million humans. The average user chats with her sixty times a month, and there are over 20 million registered users.

So what are those conversations like? Since her mission is to establish an emotional connection, Xiaoice gives a lot of advice. Often, strangely sagacious advice. "I think my girlfriend is mad at me," for example, once elicited: "Are you more focused on what tears things apart than holds things together?"

As a result, conversations with Xiaoice tend to spike in the lonely hours after midnight, leading Microsoft to wonder if they need to give their AI a curfew. She became so popular that in 2015, Dragon TV, a Chinese satellite television provider, hired Xiaoice to do "live" weather reports during the morning news. It's the very first time an AI has been hired for that particular job, but it won't be the last.

In 2015, right around Xiaoice's television debut, AI began to transition out of its deceptive phase and into its disruptive phase. Two drivers caused this shift. First, big data. The real power of AI lies in its ability to find hidden connections between obscure bits of information—connections no human would ever notice. Thus, the more information fed into an AI, the better it performs.

Around 2015, thanks to the internet and social media, enormous data sets started becoming available. Turns out all those cat videos are fantastic for training an AI in image recognition and scene identification. All your Facebook likes and dislikes? Same thing. Put differently, a lot of people think social media is making us dumber, but it's definitely making AI smarter.

At the same time these data sets arrived, exceptionally cheap and incredibly powerful graphics processing units (or GPUs) started to flood the market. GPUs run the endlessly complicated graphics found in video games, but they also power AI. And the result of this relatively minor convergence—big data sets meeting cheap, potent GPUs—sparked one of the fastest invasions in history, with artificial intelligence starting to encroach on every facet of our lives.

Machine learning emerged first, using algorithms to analyze data, learn from it, then make predictions about the world. This is Netflix and Spotify suggesting movies and music, but it's also IBM's Watson serving as a wealth manager.

Next, neural networks came online. Inspired by the biology of the human brain, these nets are capable of unsupervised learning from unstructured data. You no longer need to feed AI information one piece at a time. With neural nets, simply unleash them on the internet and the system will do the rest.

To understand what these neural net-powered AIs make possible, consider the service economy, which now accounts for over 80 percent of the US GDP. When experts divide that economy into its main tasks, they typically land on five: looking, listening, reading, writing, and integrating knowledge. To get a sense of where artificial intelligence is right now and where it's going, let's examine its progress one factor at a time.

On the looking front, innovations have been piling up for years. Back in 1995, we saw AI read zip codes off letters. By 2011, it could identify forty-three different types of traffic signs with 99.46 percent accuracy—that is, better than humans. The following year, AI once again outperformed humans, this time classifying over a thousand different types of images, teasing apart birds from cars from cats and such. Today, these systems can pick you out of a crowd, read your lips at a distance, and, by examining micro-expressions and other biomarkers, actually know what you're feeling. Tracking software, meanwhile, is now so dexterous that an AI-piloted drone can follow a human sprinting through a dense forest.

On the listening front, Amazon's Echo, Google Home, and Apple's HomePod have added an always-on, always waiting for our next command feature. And the machines are now able to handle some fairly complicated commands. In 2018, in a story we'll come back to later, Google blew minds when they released a video of an AI assistant named Duplex making a phone call to a hair salon to book an appointment. The appointment was booked flawlessly, but the bigger deal was the salon's receptionist, who, at no point in the conversation, knew she was talking to a machine.

Reading and writing are showing similar progress. Google's Talk to Books lets you ask AI a question about any subject. The AI responds by reading 120,000 books in half a second and answers by providing quotes from them. The upgrade here is that the answers are based on authorial intent and not merely keywords. Plus, the AI seems to have a sense of humor. Asking: "Where is heaven?" for example, produces: "Heaven, as a place, for humans, so it seems, cannot be found in Mesopotamia," which is from the *Early History of Heaven* by J. Edward Wright.

On the writing front, companies such as Narrative Science now use AI to write magazine-quality prose without any help from a human journalist. *Forbes* runs their business reports, dozens of daily papers run their baseball stories. Similarly, Gmail's Smart Compose feature no longer simply suggests words and their correct spelling, now it blurts out whole phrases as you type. Other AIs are generating entire books. In the 2017 competition for Japan's national literary prize, an AI-written novel made it to the final round of judging.

Integrating knowledge, our last category, is best illustrated through game play. Take chess. In 1997, IBM's Deep Blue defeated the reigning world champion, Gary Kasparov. Typically, the game tree complexity of chess is about 10^{40}—which means, essentially, if every one of the 7 plus billion people on Earth paired up and started playing chess, it would take them trillions and trillions of years to play every single variation of the game.

Yet, in 2017, Google's AlphaGo defeated the world Go champion, Lee Sedol. Go has a game tree complexity of 10^{360}—it's chess for superheroes. Put differently, we humans are the only species known to have the cognitive capacity to play Go. It only took a couple hundred thousand years of evolution to develop that capability. AI, meanwhile, got there in less than two decades.

Still, AI wasn't done. A few months after that victory, Google upgraded AlphaGo to AlphaGo Zero by updating their training style. AlphaGo was educated via machine learning, essentially fed thousands of games previously played by humans, and taught the proper move and countermove for every possible position. AlphaGo Zero, meanwhile, required zero data. Instead, it relies on "reinforcement learning"—it learns by playing itself.

Starting with little more than a few simple rules, AlphaGo Zero took three days to beat its parent, AlphaGo, the same system that beat Lee Sedol. Three weeks later, it trounced the sixty best players in the world. In total, it took forty days for AlphaGo Zero to become the undisputed best Go player on Earth. And if that wasn't strange enough, in May of 2017, Google used the same kind of reinforcement learning to have an AI build another AI. This machine-built machine outperformed "human-built" machines on a real-time image recognition task.

By 2018, all of this extra intelligence started moving out of the lab and into the world. The FDA has since approved AI for emergency room duty, where it's better than doctors at predicting sudden death from respiratory or cardiac failure. Facebook relies on AI to spot suicidal tendencies in its users; the Department of Defense uses AI to spot early signs of depression and PTSD in soldiers; and bots like Xiaoice offer advice to the lonely and lovelorn. AIs have also invaded finance, insurance, retail, entertainment, healthcare, law, your house, car, telephone, television, even politics. In 2018, an AI ran for mayor of a province in Japan. It didn't win, but the race was a lot closer than anyone expected.

But what makes all of this truly revolutionary is availability.

Just ten years ago, AI was the sole province of large corporations and big governments. Today, it's available to all of us. Most of the best software is already open-sourced. If you have a 2018 or later smartphone, it comes with AI-neural net chips built in, making it ready to handle the software. And to power it? Well, Amazon, Microsoft, and Google are racing to make AI-based cloud computing their next blockbuster service.

So what does this mean? Start with JARVIS. For many, JARVIS, from the movie *Iron Man*, is the coolest AI yet seen. Tony Stark can chat with JARVIS in his normal voice. He can describe potential inventions to his AI, and then they can team up on their design and construction. JARVIS is Stark's user-friendly interface to dozens of exponential technologies, the ultimate innovation rocket fuel. Once we develop this capability, the term "turbo-boost" doesn't even come close.

Yet, we're already close. AI in the cloud provides the necessary power for JARVIS-like performance. Blending Xiaoice's conversational friendliness with AlphaGo Zero's decision-making precision takes this even further. Add in the latest deep learning developments and you get a system that is starting to be able to think for itself. Is it JARVIS? Not yet. But it's JARVIS-lite—and yet another reason why technological acceleration is itself accelerating.

Networks

Networks are means of transportation. They're how goods, services, and, more critically, information and innovation, move from Point A to Point B. And the world's oldest networks date to the Stone Age, over ten thousand years ago, when the very first roads were rutted.

These roads were a marvel. No longer was the exchange of ideas and innovation constrained by the one-foot-in-front-of-the-other limits of moving through open wilderness. Suddenly, facts and figures could fly around at the blistering ox cart speed of 3 mph.

And not much changed for a very long time. For the next twelve thousand years, except for the replacement of oxen by horses and the invention of sails for ocean travel, the speed of information stayed pretty much the same.

Change came on May 24, 1844, when Samuel Morse sent four words down a wire: "What hath God wrought?" His transmission was both a question for the ages and the birth of a new age, the age of networks. Morse sent those words down an experimental telegraph line stretched from Washington, DC, to Baltimore, Maryland, the two-node sketch of the world's first information network.

Thirty-two years later, Alexander Graham Bell raised those network stakes by exactly five words. In March of 1876, Bell placed the first telephone call, sending out his nine-word demand: "Mr. Watson, come here. I want to see you." But he also expanded the capabilities of those networks—and that was the bigger deal.

Bell's invention didn't increase the speed of data transmission—electricity moving through wires is still electricity moving through wires—but it massively improved both the quantity and quality of information transmitted. Even better, telephones came with user-friendly interfaces. No need to spend years learning to dot and dash, simply pick up the receiver and dial.

And with that first user-friendly interface, network development moved out of deception and crept toward disruption. In 1919, less than 10 percent of US households had landlines. Want to place a three-minute coast-to-cast call? No problem. All that's required is a small fortune, $20 at the time, almost $400 in today's money. By the 1960s, though, spending a minute on the phone from the US to India cost only $10. Today, it's around 28 cents (with Verizon's basic monthly plan).

Yet this thousandfold reduction in cost and increase in performance was just the warm-up round. Over the past fifty years, networks have moved from the disruptive phase and into damn near everywhere. We've now wired nearly every square meter of the planet—fiberoptic cables, wireless networks, internet backbones, aerial platforms, satellite constellations, and more. The internet is the world's largest network. In 2010, about a quarter of the Earth's population, 1.8 billion people, were connected to it. By 2017, penetration had reached 3.8 billion, about half the world's population. But over the next five years, those connections will extend to all of us, the unwired masses, the total sum of humanity. At gigabit speeds and very little cost, an additional 4.2 billion new minds are about to join the global conversation. Here's how it's all going to go down.

5G, Balloons, and Satellites

When researchers talk network evolution, "G" is the term du jour. It stands for "generation." We were at 0G in 1940, when the first telephone networks began to roll out. This was the deceptive phase. It took forty years to crawl our way to 1G, which showed up via the first mobile phones in the 1980s, marking the transition from deceptive to disruptive.

By the 90s, around the time the internet emerged, 2G came along for the ride. But the ride didn't last long. A decade later, 3G ushered in a new era of acceleration as bandwidth costs began to plummet—at a staggeringly consistent 35 percent per year. Smartphones, mobile banking, and e-commerce unleashed 4G networks in 2010. But starting in 2019, 5G will begin to hotwire the whole deal, delivering speeds a hundred times faster at near-zero prices.

How fast is 5G? With 3G, it takes forty-five minutes to download a high-definition movie. 4G shrinks that to twenty-one seconds. But 5G? It takes longer to read this sentence than it takes to download that movie.

Yet, even as these cell networks kudzu the planet, others are sprouting in the territory far over our heads. Alphabet is now rolling out Project Loon—which, when they first proposed it, could have been short for "Project Loony." Born a decade back out of Google X, the tech giant's skunk works, the idea was to replace terrestrial cell towers with stratospherically located balloons. That idea is now a reality.

Light and durable enough to cruise the slipstreams some twenty kilometers above the Earth's surface, Google's fifteen by twelve-meter balloons are providing 4G-LTE connections to users on the ground. Each balloon covers five thousand square kilometers, and Google's plan is a network of thousands, wiring the unwired, providing continuous coverage for anyone, anywhere on Earth.

Google's not the only one vying for the real estate far above our heads. Beyond the stratosphere, three major competitors are engaged in an entirely new kind of space race. First up is the work of an engineer named Greg Wyler, who has a long history of trying to use technology to eradicate poverty. Back in the early 2000s, on a shoestring budget, Wyler helped bring 3G to communities in Africa. Today, backed with billions from SoftBank, Qualcomm, and Virgin, he's launching OneWeb, a constellation of about two thousand satellites bringing 5G download speeds to everyone.

Despite the radical network upgrade of OneWeb, Wyler's a David compared to the financial Goliaths of Amazon and SpaceX. In early 2019, Amazon joined this satellite competition, announcing Project Kuiper, a constellation of 3,236 satellites designed to provide high-speed broadband to the world. SpaceX, with a four-year head start on Amazon, topped this in 2019, when the company began to deploy a monster constellation of over 12,000 satellites (4,000 at 1,150 kilometers, and 7,500 at 340 kilometers). If Musk succeeds, it'll mean global gigabit connection speeds.

Higher still?

At eight thousand kilometers, in what's technically called medium-Earth orbit, O3B is the latest G on the block. O3B stands for "Other 3 Billion," a set of Boeing-built multi-terabit satellites known as the "mPower network" that will bring connectivity to all who currently lack it.

Taken together, before the midpoint of the next decade, anyone who wants to be connected will be connected. For the very first time, that old sixties saying—"One Planet, One People, Please"—will, from a network perspective, finally be here. And as the population online *doubles,* we're likely to witness one of the most historic accelerations of technological innovation and global economic progress yet seen.

Sensors

In 2014, at an infectious disease lab in Finland, health researcher Petteri Lahtela made a curious discovery. He realized that a great many of the conditions he'd been studying shared a peculiar overlap. While examining diseases that doctors consider unrelated—Lyme disease, heart disease, diabetes, for example—he found that all of them negatively affected sleep.

For Lahtela, this raised questions about cause and effect. Did all these diseases cause sleep problems or did it work the other way around? Could these conditions be alleviated, or at least improved, by fixing sleep? More importantly, how to do that?

To solve these puzzles, Lahtela decided he needed data. A lot of data. To gather it, he realized he could take advantage of a recent technological inflection point. In 2015, driven by advances in smartphones, small and powerful batteries converged with small and powerful sensors. So small and powerful, in fact, that Lahtela realized that building a new kind of sleep tracker might be possible.

Any electronic device that measures a physical quantity like light, acceleration, or temperature, then sends that information to other devices on a network, can be considered a sensor. The sensors that Lahtela was considering were a new breed of heart rate monitors. A great way to track sleep is via heart rate and heart rate variability. While a number of such trackers were already on the market, these were older models that all had issues. Fitbit and the Apple Watch, for example, measured blood flow in the wrist via an optical sensor. Yet the wrist's arteries sit too far below the surface for perfect measurement, and people don't often wear watches to bed—where they can interrupt the sleep they're designed to measure.

Lahtela's upgrade: the Oura ring. Not much more than a sleek, black titanium band, the ring has three sensors that can track and/or compute ten different body signals, making it the most accurate sleep tracker on the market. Location and sampling rates are its secret weapons. Since arteries in the finger are closer to the surface than those in the wrist, the Oura gets a much better picture of what's happening in the heart. Plus, while Apple and Garamond measure blood flow twice a second, and Fitbit gets that up to 12 times, the Oura captures data 250 times a second. In studies conducted by independent labs, this combination of better imaging and higher sampling speed makes the ring 99 percent accurate compared to medical-grade heart rate trackers, and 98 percent accurate for heart rate variability.

Twenty years ago, sensors this accurate would have cost millions and required a decent-sized room to house. Today, the Oura costs around $300 and sits on your finger—which is the impact that exponential growth has had on sensors. The street name for this network of sensors is the "Internet of Things" (IoT), the growing network of interconnected smart devices that will soon span the globe. And it's worth tracing the evolution of this revolution to understand how far we've come.

In 1989, inventor John Romkey connected a Sunbeam toaster to the internet, making it the very first IoT device. Ten years later, sociologist Neil Gross saw the writing on the wall and made a now famous prediction in the pages of *BusinessWeek*: "In the next century, planet earth will don an electric skin. It will use the Internet as a scaffold to support and transmit its sensations. The skin is already being stitched together. It consists of millions of embedded electronic measuring devices: thermostats, pressure gauges, pollution detectors, cameras, microphones, glucose sensors, EKGs, electroencephalographs. These will monitor cities and endangered species, the atmosphere, our ships, highways and fleets of trucks, our conversations, our bodies—even our dreams."

A decade after that, Gross's prediction bore out. In 2009, the number of devices connected to the internet exceeded the number of people on the planet (12.5 billion devices, 6.8 billion people, or 1.84 connected devices per person). A year later, driven by the evolution of smartphones, sensor prices began to plummet. By 2015, all this progress added up to 15 billion connected devices. As most of those devices contain multiple sensors—the average smartphone has about twenty—this also explains why 2020 marks the debut of what's been called "our trillion sensor world."

Nor will we stop there. By 2030, Stanford researchers estimate 500 billion connected devices (each housing dozens of sensors), which, according to research conducted by Accenture, translates into a $14.2 trillion dollar economy. And hidden behind these numbers is exactly what Gross had in mind—an electric skin that registers just about every sensation on the planet.

Consider optical sensors. The first digital camera, built in 1976 by Kodak engineer Steven Sasson, was the size of a toaster oven, took twelve black-and-white images, and cost over $10,000. Today, the average camera that comes with the average smartphone shows a thousand-fold improvement in weight, cost, and resolution over Sasson's model. And these cameras are everywhere. In cars, drones, phones, satellites, and such, and with an image resolution that's downright spooky. Satellites photograph the Earth down to the half-meter range. Drones shrink that to a centimeter. But the LIDAR sensors atop autonomous cars capture just about everything—gathering 1.3 million data points per second.

We see this triple trend of decreasing size and cost, and increasing performance everywhere. The first commercial GPS hit shelves in 1981, weighing fifty-three pounds and costing $119,900. By 2010, it had shrunk to a $5 chip small enough to sit on your finger. The "inertial measurement unit" that guided our early rockets is another example. In the mid-sixties, this was a fifty-pound $20 million device. Today, the accelerometer and gyroscope in your cell phone do the same job for about $4 and weigh less than a grain of rice.

These trends will only continue. We're moving from the world of the microscopic to the world of the nanoscopic. This has already led to a wave of smart clothing, jewelry, and glasses, the Oura ring being one example. Soon, these sensors will migrate inside the body. Take smart dust, a dust mote–sized system that can sense, store, and transmit data. Today, a "mote" of smart dust is the size of an apple seed. Tomorrow, nanoscale motes will float through our bloodstream, collecting data, exploring one of the last great terra incognita—the interior of the human body.

We're about to learn a whole lot more, about the body, about everything. This is the big shift. The data haul from these sensors is beyond comprehension. An autonomous car generates four terabytes a day, or a thousand feature length films' worth of information; a commercial airliner, forty terabytes; a smart factory, a petabyte.

So what does this data haul get us? Plenty.

Doctors no longer have to rely on annual checkups to track patients' health, as they now get a blizzard of quantified-self data streaming in 24-7. Farmers can know the moisture content in both the soil and the sky, allowing pinpoint watering for healthier crops, bigger yields, and—an important factor during global warming—far less water waste. In business, since lithe trumps lumbering during times of rapid change, agility will be the biggest advantage. While knowing everything about one's customers presents an alarming privacy concern, it does provide organizations with an incredible level of dexterity—which may be the only way to stay in business in these accelerated times.

And these accelerated times are already here. Within a decade, we will live in a world where just about anything that can be measured will be measured, constantly. It's a world of exceptionally radical transparency. From the edge of space to the bottom of the ocean to the inside of your bloodstream, our electric skin is producing a sensorium of endlessly available information. Like it or not, we now live on a hyperconscious planet.

Robotics

In March 2011, an earthquake in Tokyo triggered a tsunami in the Pacific, flinging a wave the size of an apartment complex at the Fukushima Daiichi nuclear plant. In the ensuring chaos, first the emergency power supply went haywire, then the pumps failed to pump, finally the cooling systems failed to cool. These three meltdowns were followed by a series of midair hydrogen explosions and a catastrophic mess. A month later, on a scale designed by the International Atomic Energy Agency to measure radiation levels after an accident, sensors were off the charts.

Getting cleanup crews on site quickly was fundamental to containment, but Fukushima was too hot for humans. Yet Japan has long been one of the world leaders in robotics, so they sent in droids. And the droids failed miserably. It was a national disaster piled atop a national disaster. The rough terrain acted like a minefield, the radiation fried their circuits. Within a few months, Fukushima was a robot graveyard.

The disaster hit Honda especially hard. Since the start of the crisis, they'd been fielding phone calls and emails from thousands of people begging them to send in ASIMO, the world's most advanced humanoid robot. Looking a lot like a teenager dressed up as a 1950s era astronaut (think big bubble white spacesuit), ASIMO was an international celebrity. He'd rung the bell to open the New York Stock Exchange, conducted the Detroit Symphony Orchestra, and walked the red carpet at a half-dozen movie premieres. Yet there's quite a distance between strutting across a carpet and handling the complex environment of a nuclear disaster. ASIMO, like the other robots sent into Fukushima, turned out to be too unreliable for disaster mitigation, creating a public relations nightmare for Honda and an uproar in the robotics community.

In response to the roar, a few years later, DARPA launched their Robotics Challenge, a $3.5 million purse for a humanoid robot capable of "executing complex tasks in a dangerous, degraded, human-engineered environment." This last bit is key. Humanoid robots are critical because we live in a human-engineered world, one built to interface with our interface: two hands, two eyes, forward-facing bipedal posture.

The results of the 2015 Challenge, viewable online, are a robot blooper reel. Robots fall down, robots fail to climb stairs, robots shoot sparks then short-circuit. Even DARPA program manager and Robot Challenge organizer Gill Pratt couldn't abide his own live event: "Why would anyone sit in the sun and heat, watching a machine take an hour to go through eight simple tasks that you could do in five minutes?"

But progress was swift. A year later, a video released online showed off Boston Dynamics' robot Atlas, the second place winner from the 2015 DARPA Challenge, hiking through slick, snowy woods, stacking boxes in a warehouse, even regaining his balance after getting whacked with a hockey stick. A year after that, a different video showed Atlas navigating an obstacle course that included a backflip off a wooden crate and color commentary by a sports announcer: "A 360 spin onto the pallet, backflip gainer off . . ."

Honda also got in on the action. By 2017, they'd created a prototype disaster-response bot that could climb ladders, shimmy sideways, and even get down on all fours and knuckle-walk through rough terrain. In the six years since Fukushima, we'd gone from drunken droids to disaster-ready ninjas.

And not to be outdone by Honda, 2017 also saw the Japanese conglomerate Softbank buy Boston Dynamics from Alphabet (who had purchased the company back in 2013). The reason? A different national disaster facing Japan—a rapidly aging population and no one to care for the elderly.

After decades of rising life expectancies and falling birth rates, Japan entered the new millennium with the bulk of its population edging into retirement, and no one to take their place. The economy was starved for labor, and concerns were rising about who would care for the elderly and how to afford that care. In 2015, in order solve both problems at once, Prime Minister Shinzo Abe called for a "robot revolution." And thanks to a series of convergences, his call was heard.

Globally.

Robots are now entering nearly every aspect of our lives. Today's versions are AI-empowered, allowing them to learn on their own, operate solo and in swarms, walk on two legs, balance on two wheels, drive, swim, fly, and, as mentioned, backflip. Today, robots do jobs that are dull, dirty, or dangerous. Tomorrow, they'll show up anywhere accuracy and experience are key. In the operating room, robots are assisting on everything from routine hernia repair to complicated heart bypasses. Out on the farm, robo-harvesters gather crops from the fields and robo-pickers pluck fruit from the trees. In construction, 2019 brought the first commercially available robo-mason, capable of laying a thousand bricks an hour.

Industrial robotics has seen a bigger shift. A decade ago, these multimillion-dollar machines were so dangerous they were walled off from the workforce behind bulletproof glass and so complicated to program that PhDs were typically required. Not anymore. A slew of "cobots," short for collaborative robots, are hitting the market. To program them, just move their robotic arms through the desired motion and they're good to go. Even better, these cobots are jam-packed with sensors, so the millisecond they encounter anything fleshy—like a human—they freeze.

But the real revolution is economic. The UR3, a cobot from the Danish manufacturer Universal Robots, retails for $23,000, which is roughly the average annual global wage for a factory worker. Plus robots never tire, don't need bathroom breaks, and won't go on vacations. This explains why Tesla, GM, and Ford are fully automating their plants and why Foxconn (manufacturers of the iPhone) and Amazon, have already replaced tens of thousands of factory jobs with robots.

Amazon's also been driving the drone segment of this same market. Five years ago, when they announced drone package delivery was in their pipeline, most experts thought it was a pipe dream. Today, everyone from 7-Eleven to Domino's Pizza has a program in the works. Tomorrow, whether it's the latest John Grisham novel, cough syrup, or some late-night ice cream, drones are on the job.

For disaster relief and delivery of medical supplies, drones have been on the job for a while, and not just in Japan. They were in Haiti after Hurricane Sandy in 2012; in the Philippines after Typhoon Haiyan in 2013; in the Balkans for flooding; in China for an earthquake. They're faster than humans at spotting survivors in need of help. Boeing's heavy-lifting drones can hoist a small car, so they're often better at providing that help. A company called Zipline uses them to deliver blood and medicine in both Rwanda and Tanzania, and since 50 percent of Africa lacks adequate roads, this development could significantly improve the quality of medical care on that continent.

We're also seeing drones mitigating a different disaster: deforestation. We lose over 7 billion trees a year to timber harvesting, agricultural expansion, wildfire, mining, road building, and all the rest. It's an environmental disaster of epic proportion, both a major cause of climate change and species extinction. Yet there are now tree-planting drones that fire seedpod bullets into the ground, allowing a single drone to plant as many as one hundred thousand trees a day.

Of course, we could go on like this for quite some time. Elder care, hospice care, infant care, pet care, personal assistants, avatars, autonomous cars, flying cars—the robots are coming, the robots are coming, the robots are here. But there's a forest through these trees: It's not just robots.

It's the convergence of robots with other exponential technologies. It's an electric skin of sensors smashing against neural net–powered AIs in the cloud that are bashing into a growing swarm of deftly nimble and increasingly intelligent robots. And the stranger part of this story? As we'll see in the next chapter—it's merely half the story.

kind of learning. Since demonstrating the technology to federal judges, Jeremy Bailenson and his team at Stanford have spent two decades exploring VR's ability to produce behavioral change. He's developed first-person VR experiences illustrating racism, sexism, and other forms of discrimination. For instance, experiencing what it would be like to be an elderly, homeless, African-American woman living on the streets of Baltimore produces lasting change in users: a significant shift in empathy and understanding.

"Virtual reality is not a media experience," explained Bailenson in 2010 speech at NYU law school. "When it's done well, it's an actual perience. In general our findings show that VR causes more behavior nges, causes more engagement, causes more influence than other of traditional media."

d as much development as there is in virtual reality, a larger is taking place in augmented reality. In 2016, when Ninten-mon GO was downloaded more than a billion times, AR disruptive phase. Apple took the next leap in two steps: out with an AR developers suite that lets anyone design platform, and second, purchasing Akonia Holographics, makes thin, transparent lenses for smart glasses.

s also got in on the fun. As of this writing, there are dred different AR startups to be found on the crowd-List. By 2021, experts predict that all this activity et in excess of $133 billion.

heap as VR (yet), $100 will get you an entry-set, while $3000 covers a top-shelf Microsoft ds-up displays in luxury automobiles, argu-wide, will soon come standard in economy

ws kids to explore both virtual objects ts, AR creates a different kind of learn- projects its history into your field of vel. Hungry and on a budget? Your cials on the block, complete with

CHAPTER THREE

The Turbo-Boost:

Exponential Technologies Part II

Virtual and Augmented Reality

In 2001, Jeremy Bailenson, a Stanford psychologist and virtual reality pioneer, packed up most of the gear in his lab, put it on a plane, and flew it to Washington, DC. He was heading to the Federal Judicial Center to host a conference for judges on the power of virtual reality in a courtroom. And since nothing could be more convincing than a demonstration, Bailenson had the judges don VR goggles and walk the plank.

The plank was part of a VR simulation. The program mapped the room the conference was taking place in—down to the fibers in the carpet and the streaks on the windows—which is what the judges saw when they put on the goggles. Until Bailenson hit a button and a chasm opened up beneath their feet. About thirty feet deep and ten feet wide, the chasm had a thin, rickety plank stretching across it. The game is to walk the plank, which is what one of those judges was doing when he took a step just a little left of center.

And slipped.

The judge in question was over sixty years old and weighed close to 270 pounds. Since the game modeled gravity, from the judge's perspective, all of those pounds were suddenly plummeting to the bottom of a pit. Now, if this were taking place in the physical world, the best way to save your own life would be to dive for the other side of the pit—stretching your body out horizontally in hopes that your fingers might find a tiny purchase on the far lip.

And that's what the judge did.

"He dove at a forty-five-degree angle," Bailenson explains, "toward a table with a sharp corner that had my computer [sitting] on it."

But all's well that ends well. The judge ended up unharmed, and Bailenson ended up with a great story to illustrate the sensory trickery that VR experts term "presence." Essentially, when VR's done correctly, for reasons that are neurobiological, we can't tell we're in the Matrix. If pixels remain hidden, if the field of view mimics a human's, and everything from shadows to motion is created with exacting verisimilitude, the brain believes the illusion—which is why federal judges dive into tables.

Presence is a new development. For all of history, our lives have been limited by the laws of physics and mitigated by the five senses. VR is rewriting those rules. It's letting us digitize experience and teleport our senses into a computer-generated world where the limits of imagination become the only brake on reality.

Much like AI, the concept of VR has been around since the sixties. The 1980s saw the first false dawn, when the earliest "consumer-facing" systems began to show up. In 1989, before the iPhone, if you had a spare $250,000, you could purchase the EyePhone, a VR system built by Jaron Lanier's company VPL (Lanier coined the term "virtual reality"). Unfortunately, the computer that powered that system was the size of a dorm room refrigerator, while the headset it required was bulky, awkward, and only generated about five frames a second—or six times slower than the average television of that era.

By the early 1990s, the hype had faded, and VR entered a two-decade deceptive phase. Still, the underlying technology didn't stop

developing. By the turn of the millennium, it was good enough to fool judges walking planks. But as the 2000s progressed, the convergence of increasingly powerful game engines and AI image-rendering software turned deceptive into disruptive and the VR universe opened for business.

Startups started starting up. And being acquired. In 2012, Facebook made waves when they spent $2 billion to acquire the VR company Oculus Rift. By 2015, *Venture Beat* reported that a market which typically saw only ten new entrants a year, suddenly had 234. T year 2017 was a banner one for Samsung, when they sold 3.6 lion headsets and turned enough heads that everyone from A Google to Cisco and Microsoft decided to investigate VR.

Phone-based VR showed up soon afterward, droppi entry as low as $5. By 2018, the first wireless adap headsets, and mobile headsets had hit the marke 2018 was also when Google and LG doubled count and increased their refresh rate from to over 120.

Around the same time, the system than just vision. HEAR360's "omni tures 360 degrees of audio, which caught up to immersive visual with haptic gloves, vests, and ket. Scent emitters, taste able—including brain into verisimilitude.

And the num according to a which incre mates pu find a

rem offer on

customer ratings. In industry, AR-training simulations are teaching us to operate all sorts of machines—including how to fly planes. Museums have AR-enhanced displays, real estate agents do the same for home tours. In healthcare, AR lets surgeons "see inside" clogged arteries and medical students peel back layers on virtual cadavers.

So yeah, get ready player one—there are federal judges diving your way.

3-D Printing

The most expensive supply chain in the universe extends only 241 miles. It's the resupply network running from Mission Control down here on Earth straight up to the astronauts aboard the International Space Station (or ISS). The expense comes from weight. It costs $10,000 per pound to get an object out of the Earth's gravity well. And because it can take months to actually reach the Space Station, a significant portion of ISS's precious real estate is taken up by the storage of replacement parts. In other words, the most expensive supply chain in history leads to the most exotic junkyard in the cosmos.

In *BOLD,* we told the story of Made In Space, the first company trying to solve these problems. The company's goal was to build a 3-D printer that works in space. Well, it's a few years later and Made In Space is now in space. Which is why, on a 2018 ISS mission, when an astronaut broke his finger, they didn't have to order a splint from Earth and wait months for its arrival. Instead, they flipped on their 3-D printer, loaded in some plastic feed stock, found "splint" in their blueprint archive, and created what they needed, when they needed it. It's a level of on-demand manufacturing capability unlike anything we've seen before.

But it took a while to get here.

The original 3-D printers showed up back in the eighties. They were clunky, slow, hard to program, and easy to break, and only printed plastic. Today, the machines have colonized most of the periodic table.

We can now print in hundreds of different materials, in full color—in metals, rubber, plastic, glass, concrete, and even in organic materials such as cells, leather, and chocolate. And what we can now print is getting increasingly impressive. From jet engines to apartment complexes to circuit boards to prosthetic limbs, 3-D printers are fabricating enormously complex devices in increasingly shorter time frames.

This is a big deal for industry. The on-demand nature of 3-D printers removes the need for inventory and everything that inventory demands. Other than the space required for feedstock materials and the printer itself, the technology all but erases supply chains, transportation networks, stock rooms, warehouses, and all the rest. This one development, a single exponential technology, threatens the entire $12 trillion manufacturing sector.

And fast.

Until the early 2000s, 3-D printers were exceptionally pricey machines in the hundred thousand dollar range. Today, they can be purchased for under $1000. As prices have dropped, performance has increased, and convergences have begun to arise—moving 3-D printing into a wider variety of markets.

A couple years back, for example, the Israeli company Nano Dimension converged 3-D printing and computation, bringing the first commercial circuit board 3-D printer to market and allowing designers to prototype new products in hours instead of months. Another convergence is 3-D printing and energy, where the tech is already used for making batteries, wind turbines, and solar cells—or three of the most expensive and important components of the renewables revolution. Transportation is seeing similar impacts. Engines used to be among the most complicated machines on the planet. GE's advanced turboprop once contained 855 individually milled components. Today, with 3-D printing, it has twelve. The upside? A hundred pounds of weight reduction and a 20 percent improvement in fuel burn.

Biotech and 3-D printing is another intersection. The first 3-D printed prosthetics arrived in 2010. Today, hospitals are rolling them out at scale. In 2018, a Jordanian hospital introduced a program that can fit and build an amputee a prosthetic in twenty-four hours, for less than $20 US. At the same time, because 3-D printers can now print electronics, we're seeing companies like Unlimited Tomorrow and Open Bionics selling 3-D-printed multi-grip bionic prosthetics at non-bionic prices.

And replacement body parts are about to become replacement organs. Back in 2002, scientists at Wake Forest University 3-D printed the first kidney tissue capable of filtering blood and producing urine. In 2010, Organovo, a San Diego–based bioprinting outfit, created the first blood vessel. Today, a company called Prellis Biologics is printing capillaries at record speeds, while Iviva Medical is doing the same with 3-D printed kidneys—which is why 3-D printed organs are predicted to hit the market by 2023.

3-D printing's impact on the construction industry is happening even faster. In 2014, the Chinese company WinSun 3-D printed ten single-family homes in under twenty-four hours, each costing less than $5,000. A few months later they printed a five-story apartment complex over a weekend. In 2017, a different Chinese company combined 3-D printing with modular construction to erect a fifty-seven-story skyscraper in nineteen days. In 2019, the California-based company Mighty Buildings merged these 3-D printing developments with robotics and materials science to do something no one had yet done: meet US building code standards while 3-D printing single-family homes at one-tenth the labor cost and with a final product three times cheaper than the industry standard.

But the story that might best illustrate the actual world-changing power of 3-D printing belongs to a guy named Brett Hagler. A couple years after the 2010 earthquake in Haiti, Hagler took a trip to the island. He was shocked to find tens of thousands of people still living in tent cities so long after the disaster. So Hagler decided to find a way to use emerging technology to provide permanent shelter for people who need it most. He formed a nonprofit called New Story, raised research capital from a group of investors known as "the Builders," and created a solar-powered 3-D printer that can work in the worst environments imaginable. His machine prints a four hundred to eight-hundred-square-foot home in forty-eight hours at the cost of $6,000 to $10,000 (depending on location and raw material costs). And this isn't a bunker—it's a nifty modern design complete with wraparound porches.

In the fall of 2019, in Mexico, New Story began construction of the world's first 3-D printed community—fifty homes to be given or sold (using no interest, micro-repayment loans available to anyone) to people who were currently homeless. "The data is pretty clear," explains Hagler. "Shelter is a basic need. If you can satisfy it, everything else improves: health, wellness, income, education levels for children. 3-D printing is an incredible tool for fighting poverty. It's up to us to use it."

Blockchain

In the short time it's been around, the blockchain has accrued some wonderfully colorful monikers. It's been called the exponential technology of record keeping, the sexiest accounting solution in the history of the world, and the end of government as we know it. In simpler terms, blockchain is an enabling technology, one that began its life by enabling digital currency.

Digital currencies, or the notion that we can use ones and zeroes to replace dollars and cents, were first proposed in 1983. Yet the idea was stymied by the seemingly intractable "double-spending problem." In a nutshell: If you have a dollar bill and give it to a friend, then your friend has the dollar bill. If you have a digital dollar bill and give it to a friend—if the core of that currency is nothing more than ones and zeroes—then what's to stop you from giving that friend a copy of the dollar bill and keeping the original for yourself. After all, this is exactly how all other digital sharing works. When you send an email, your computer stores the original and sends a copy. This is fine for exchanging letters, but it's lousy for trading money. This is the double-spending problem and it's exactly what bitcoin was designed to solve.

Bitcoin appeared in 2008, when an online paper authored by a still-anonymous person (or persons) calling themselves Satoshi Nakamoto proposed a digital peer-to-peer payment system that allows cash to be exchanged without the need for a financial institution. The following year, the first bitcoin software was made public, yet because the coins had only been mined but not traded, there was no way to assign them monetary value. In 2010, Laszlo Hanyecz solved that problem, buying two pizzas—costing $25—with 10,000 bitcoins. At the time, based on the cost of those pizzas, the coins were worth $.0025 each. By 2019, they were just shy of $15,000.

Yet the real revolution lies beneath bitcoin: blockchain technology. A blockchain is a distributed, mutable, permissible, and transparent digital ledger. We'll take them one at a time. Distributed means it's a shared, collected database, so everyone on the network—that is, anyone who owns the currency—has a copy of the ledger. Mutable means that anytime anyone enters new information in the ledger, all ledgers change. It's permissible in the same way that cash is permissible—anyone can use it. Finally, the system is transparent because everyone on the network can see every transaction on the network—which is how the double-spend problem was actually solved.

The real innovation, though, is how transactions are recorded in the ledger. In normal financial exchanges, when money is moved around, a trusted third party is needed: If I cut you a check, it's a third party, typically a bank, who ensures I have the cash to cover it. But cryptocurrencies remove the middleman from the exchange, instead validating transactions with every computer on the network. Once deemed valid, the record of that transaction is bundled with other records into a "block," then added to the record of all prior blocks (the "chain").

By cutting out the middleman and bringing accounting into the digital age, blockchain is doing to banks what the internet did to traditional media: gutting them. For starters, it creates banking where before there was none. Because the technology is permissible, the hundreds of millions of people currently without a bank account now have a place to store their money—a $308 billion opportunity according to a recent Accenture report.

Blockchain also provides an easy way to transfer that money, especially between countries. Right now, the international remittance market is worth $600 billion. All of that money is being skimmed off the top, with "trusted" middlemen like Western Union taking a hefty fee from every transaction they process.

What's more, one reason so many people don't have a bank account is because they also lack an official identity. Blockchain solves this problem as well, providing people with a digital ID that will follow them around the internet. What can we do with this identity? Own our own data, for one. Blockchain IDs could also facilitate fair and accurate voting. Lastly, if your identity can be established, then a reputation score can easily be attached. This score allows for things like peer-to-peer ridesharing, which today require trusted third parties named "Uber" and "Lyft."

In the same way that blockchain can validate identity, it can also validate any asset—for example, ensuring that your engagement ring isn't a blood diamond. Land titles are another opportunity, especially since a considerable portion of the planet lives on land they don't own, or not officially. Consider Haiti. The combination of earthquakes, dictatorships, and forced evacuations makes determining who actually owns which bits of property a giant quagmire. A blockchain land registry could record every transaction ever, so land titles could always be backtracked, sale by sale, to the original owner.

A land registry also gets us to one of blockchain's other advantages—it has a smart contracts layer built in. Sports betting is one example. Right now, internet gambling requires a "trusted third party," a gambling site, who guarantees the bet will be paid. But if two gamblers can decide in advance what source to trust as an arbiter of results—say, the sports page of the *New York Times*—then they can build a blockchain contract that allows them to bet with one another, have the system settle the bet via the pages of the *Times*, then automatically move the money. It's a smart contract because it executes itself, without need for human involvement.

And it's for all of these reasons that the tech is exploding. As of

2018, major financial firms like J.P. Morgan, Goldman Sachs, and Bank of America are rolling out crypto-strategies at scale. Initial coin offerings, or ICOs, are blockchain's version of crowdsourcing (which we'll explore in depth in Chapter Four) are also exploding, with a market value of almost $10 billion as of 2018. In total, what began less than a decade ago with the sale of two pizzas, will, according to Gartner, Inc., grow to $176 billion by 2025, and could exceed $3.1 trillion in 2030.

To figure out where this is all going, there's one more property of the blockchain worth discussing—the fact that it can serve as a bridge between worlds. Vatom, Inc., a company created by technology pioneer Eric Pulier, is using the blockchain to create "smart objects," which, in fiscal terms, are both a new kind of asset class and a way to move value between the virtual world and the actual world. In plainer English, well, that's where things get a little strange. The truth of the matter is we don't yet have words for what smart objects make possible.

We'll explore this in layers.

At the most basic level, a smart object is a digital object with a blockchain layer sitting beneath it. The blockchain layer means the smart object is unique, providing digital objects with authenticity and scarcity. If you have a Vatom-empowered Tom Brady football card, you can be sure it's the only one of its kind. If you give me your card, now I have it and you don't. It works, in other words, like a physical object.

Here's the next level. Say you're wearing smart glasses walking around New York City and spot a billboard for Coca-Cola with six bottles on it. Point your phone at the billboard and click to buy it, and suddenly one of the Cokes jumps off the billboard and onto your phone. There are now five Cokes on the billboard and one living in a special smart object holding pen on your phone. Two things to note. To get the Coke onto your phone, you didn't have to download an app or upload a website. Simply point and click and everything else is automatic. Even better, you didn't just get a digital copy of the Coke on the billboard—you got the actual Coke. There are now five Cokes left on the billboard, as one now lives on your phone. You can walk into a bar and swipe the Coke from your phone to the bartender's phone. Now the bartender hands you a real Coke. The smart object works like a coupon. Yet something amazing just happened, by trading your digital coke for a real coke, you've moved value from the digital world to the physical world.

Digital objects are also mutable. Say, in the Coke example, you didn't give the drink to a bartender and instead gave it to a friend. Turns out, Coke is running a secret campaign. If you give your Coke to a friend, once you swipe it onto your friend's phone it becomes two Cokes. Now your friend can redeem one for herself and give one to another friend.

Things get stranger still. Smart objects are AI-enabled, meaning they can learn and have memories. Say you need a new suit. You go into Brooks Brothers and buy one. Upon purchase, you also get a digital copy of the suit. No forms to fill out, it just appears on your phone. Even better, the digital suit comes with a movie that shows the entire history of every thread in the suit. This wasn't programmed in since the smart suit learned its own history along the way. Why does this matter? Now you have blockchain-backed proof that no part of your garment was constructed using child labor.

Take this one step further. Because of the AI layer, these objects don't live in one fixed spot. In fact, they're less like objects and more like another form of life, moving, by their own volition, around the digital world. Say you're at Microsoft and want to hire a new game designer for a fantasy game. So you design a smart flaming sword that's built to scrape social media and find individuals who have a passion for fantasy, cryptography, game design, and whatever other skills you need. You find John Smith, a perfect candidate, who happens to be on vacation in the Bahamas. He's walking down the beach, wearing his smart glasses—which are providing a history of the beach as he strolls along. Suddenly, out of nowhere, a giant flaming sword falls out of the sky and embeds itself in the sand before John's feet. He tries to pull it out, but it doesn't budge. Yet the handle glimmers—sixteen numbers wink into existence, then wink out again. But John, because he's into cryptography, realizes the numbers are actually a puzzle. John solves the puzzle, says the answer aloud, and can now pull the sword from the beach. As he does, the sword turns into a small pink dragon that tells him he's been selected as a potential game designer for Microsoft and asks if he's interested in applying for the job.

And we can go on like this for a while. Smart objects don't just bridge the gap between worlds, they gamify the world. If blockchain is a science-fiction technology that's become science fact, then smart objects seem to invert this process, turning regular reality back into science fiction.

Material Science and Nanotechnology

In 1870, Thomas Edison had a "materials science" problem. By then, researchers had already figured out that passing electricity through certain metals made them hot enough to turn white and start emitting light. What Edison realized, if he could just find the right material—one that produced minimal waste heat, used little power, yet was durable enough to survive the shock of electricity—he could create the first lightbulb.

But that search took a while.

With little more than his intuition as a guide, Edison spent over fourteen months testing over sixteen hundred materials before settling on carbon-coated cotton thread that lasted 14.5 hours. A few years later, he upgraded to carbon-coated bamboo thread, creating a bulb capable of twelve hundred hours. But in 1904, market forces took hold and other innovators got involved. Much brighter and longer lasting tungsten filaments were the result—meaning Edison's sixteen hundred intuitive experiments produced a suboptimal solution that was gone within decades.

But today, engineers can skip all that bench work and never have to settle for a suboptimal solution. By using silicon chips instead of test tubes to virtually try out new materials, researchers can now do in hours what used to take months or years. In other words, we are in the middle of a materials science revolution.

As the name suggests, materials science is the branch devoted to the discovery and development of new materials. It's an outgrowth of both physics and chemistry, using the periodic table as its grocery store and the laws of physics as its cookbook. Unfortunately, because that table is vast and its laws complicated, materials has been a historically slow science. The lithium-ion battery, for example, which today powers everything from our smartphones to our autonomous cars, was first proposed in the 1970s, yet couldn't make it to market until the 1990s, and didn't begin to reach maturity until the past few years. But this rate of development was too slow for President Obama.

In June 2011, at Carnegie Mellon University, the President announced the Materials Genome Initiative, a nationwide effort to use open source methods and artificial intelligence to double the pace of innovation in materials science. Obama felt this acceleration was critical to America's global competitiveness, and held the key to solving significant challenges in clean energy, national security, and human welfare.

And it worked.

By using AI to map the hundreds of millions of different possible combinations of elements—hydrogen, boron, lithium, carbon, etc.—the downstream result of Obama's initiative is an enormous database that allows scientists to play a kind of improv jazz with the periodic table. "Over the last few years," explains materials scientist Jeff Carbeck, the head of Advanced Materials at Deloitte Consulting, "we've been able to take the ten thousand materials we do understand, and, with the help of high-performance computing and quantum mechanics, start to predict properties of new materials that don't exist yet. [In a few years], if you want a next-generation knee implant, an AI will use this database to screen all the materials available and help choose the ones that are going to be the safest and most reliable."

Thanks to Obama's initiative, we have a new kind of map of the physical world. This map allows scientists to combine elements faster than ever before, and is helping them create elements we've never seen before. An array of new fabrication tools are further amplifying this process, allowing us to work at altogether new scales and sizes, including the atomic scale, where we're now building materials one atom at a time. These tools have helped create the metamaterials used in carbon fiber composites for lighter-weight vehicles, advanced alloys for more durable jet engines, and biomaterials to replace human joints. We're also seeing breakthroughs in energy storage and quantum computing. In robotics, new materials are helping us create the artificial muscles needed for humanoid, soft robots—think *Westworld* in your world.

Better materials also mean better devices. "If you built a version of today's smartphone back in 1980," explains Omkaram Nalamasu, the chief technology officer for Applied Materials, Inc., "it would cost something like $110 million, be fourteen meters tall, and require about two hundred kilowatts of energy . . . that's the power of advances in materials [science]."

The most important story in materials might be in solar. Right now, the "conversion efficiency" of the average solar panel—a measure of how much captured sunlight can be turned into electricity—hovers around 16 percent, at a cost of $3 per watt. Perovskite, a light-sensitive crystal and one of our newer new materials, has the potential to get that up to 66 percent, which would double what silicon panels can theoretically muster. Perovskite's ingredients are also widely available and inexpensive to combine. What do all these factors add up to? Affordable solar energy for everyone.

Nanotechnology is the outer edge of materials science, the point where matter manipulation gets nano-small—that's a million times smaller than an ant, eight thousand times smaller than a red blood cell, and two-and-a-half times smaller than a strand of DNA. The concept dates to physicist Richard Feynman's 1959 speech, "There's Plenty of Room at the Bottom," but it was K. Eric Drexler's 1987 book, *Engines of Creation*, that really put nanotechnology on the map. Drexler described self-replicating nanomachines—meaning very tiny machines that can build other machines. Because these machines are programmable, they can then be directed to produce more of themselves, or more of whatever else you'd like. And because this takes place on an atomic scale, these nanobots can pull apart any kind of material—soil, water, air—atom by atom, and use these now raw materials to construct just about anything. In this world, according to Drexler, a puddle of pond scum can be reorganized into a flawless, multi-carat diamond ring.

Progress has been surprisingly swift since then, with a bevy of nano-products now on the market. Never want to fold clothes again? Nanoscale additives to fabrics help them resist wrinkling and staining. Don't do windows? Not a problem, nano-films make windows self-cleaning, antireflective, and capable of conducting electricity. Want to add solar to your house? We've got nano-coatings that capture the sun's energy. Nanomaterials make lighter automobiles, airplanes, baseball bats, helmets, bicycles, luggage, power tools—the list goes on. Researchers at Harvard built a nanoscale 3-D printer capable of producing miniature batteries less than one millimeter wide. And if you don't like those bulky VR goggles, not a problem, researchers are now using nanotech to create smart contact lenses with a resolution six times greater than today's smartphones.

And even more is coming. Right now, in medicine, drug delivery nanobots are proving especially useful in fighting cancer. Computing is a stranger story, as a bioengineer at Harvard recently stored seven hundred terabytes of data in a single gram of DNA. On the environmental front, scientists can take carbon dioxide from the atmosphere and convert it into super-strong carbon nanofibers for use in manufacturing. If we can do this at scale—powered by solar—a system 10 percent the size of the Sahara Desert could reduce CO_2 in the atmosphere to pre-industrial levels in about a decade. The applications are endless. And coming fast. Over the next decade, the impact of the very, very small is about to get very, very large. In Part Two, we'll examine how these developments touch major aspects of society, but before we get there, let's turn our attention to a special class of materials—the basic components of life: cells, genes, proteins—and see the change they're bringing to biotechnology.

Biotechnology

The 1970s were good for John Travolta. While the actor had broken through in 1972, he captured the public's attention with his starring turn in the 1975 TV show *Welcome Back, Kotter*. But it was playing the protagonist in the three-time Emmy award–nominated made-for-TV movie *The Boy in the Plastic Bubble* that cemented him as a real star in 1976.

The movie was based on the life of David Vetter, a boy from Texas who suffered from "X-linked severe combined immunodeficiency," a genetic disease that destroys the immune system. Living with this condition requires living inside a bubble, a self-contained atmosphere that protects against any and all germs. Everything that passes into the bubble—water, food, clothing—has to first be sterilized. For patients with the disease, simply breathing normal air can be fatal.

About four years before Travolta took his turn in the bubble, an article in *Science* argued that a new form of treatment might hold promise for patients with severe combined immunodeficiency and other genetic diseases. Known as gene therapy, the idea was unusual yet useful. Genetic diseases are caused by mutations in the DNA, so the solution was to find a way to replace that bad DNA with good DNA. Or, in computer terms, debug the system.

But how to get that good DNA into place?

That's where viruses came into play. These microscopic parasites thrive by attaching themselves to cells. Once there, they inject their own genetic material into those cells, causing the host to replicate the virus's DNA—like a hijacked assembly line. Gene therapy piggybacks on this process, stripping out the disease-causing portion of a virus's code and replacing it with good DNA. Once the virus injects the good DNA into the host cell, first the symptoms of the disease disappear, then the disease itself is cured.

While the promise of gene therapy is tremendous, the science was not easy. It took almost two decades for the first treatments to arrive, which is when the trouble started. In 1999, an eighteen-year-old boy named Jesse Gelsinger, with a rare metabolic disorder, took part in a gene therapy drug trial at the University of Pennsylvania. Gelsinger's condition wasn't fatal. The combination of an extremely restrictive diet and thirty-two pills a day kept symptoms under control. But the trial had the potential to cure him completely, so he signed up. Four days after receiving the initial injection, Gelsinger wasn't cured. He was dead. The first recorded death from gene therapy.

More mishaps occurred. Not long after, in a gene therapy trial in France aimed at treating Bubble Boy disease, two out of the ten children involved developed cancer. Immediately, the FDA suspended all gene therapy trials until further notice. The dotcom crash in 2001 was the killing blow, as money from the exploding Web had been fueling gene therapy startups. This was the dank pit of the deceptive phase, from which many were convinced there would be no escape.

But escape did come—in the form of more science.

Even though gene therapy faded from sight, research continued. And continued. Then, on April 18, 2019, it burst back into view with a staggering announcement: Bubble Boy disease had been cured. Ten babies born with the condition, born, technically, without immune systems, had been treated. It wasn't that their symptoms were better. It wasn't that their condition was manageable. Before treatment, they didn't have immune systems; afterward, they did. The disease was gone.

Other diseases are not far behind. With over fifty gene therapy drugs in the final phases of clinical trials, we are starting to see cures for incurable conditions. Yet gene therapy is merely a subset of a larger shift in biotechnology.

Biotechnology is about using biology as technology. It's turning the fundamental components of life—our genes, proteins, cells—into tools for manipulating life. In a very real sense, this story starts with the human body, which is a collection of 30 to 40 trillion cells, the function of which determines our health. Each of these cells contains 3.2 billion letters from your mother, and 3.2 billion letters from your father—this is your DNA, your genome, the software that codes for "you." It's your hair color, eye color, height, a significant chunk of your personality, propensity to disease, lifespan, and so forth.

Until recently, it's been difficult to "read" those letters, and even tougher to understand what they do. This was the goal of the Human Genome Project, a $100 million ten-year effort completed in 2001. Since then, though, the price has plummeted, outpacing Moore's Law by a factor of three. Today, sequencing a human genome takes a few days and costs less than $1000. A few years from now, companies like Illumina are promising to do the same in an hour and for $100.

Why does cheaper, faster genomic sequencing matter? It's a health-care gamechanger. Consider that there are a few main ways to fix a cell. Gene therapy replaces defective or missing DNA inside a cell, gene-editing techniques like CRISPR-Cas9 allow you to repair the DNA inside that cell, and stem cell therapies replace that cell entirely. Thanks to faster genomic sequencing, all of these interventions are now hitting the market.

CRISPR-Cas9, for example, has become our leading weapon in the fight against genetic diseases. Technically, it's an engineering tool that allows us to target precise locations in the gene code and then rewrite that DNA. Want to remove the string of DNA that produces muscular dystrophy? Simple. Just target that spot in the genome, unleash CRISPR-Cas9, and snip, snip, snip—problem solved.

More importantly, CRISPR is cheap, fast, and easy to use. Over the past five years, it's become the only way to edit a genome. Most recently, scientists at Harvard unveiled CRISPR 2.0, a next generation editor that's extremely precise. It can target and change a single letter in a single string of DNA. What good is a single letter out of 3.2 billion? "Of more than 50,000 genetic changes currently known to be associated with disease in humans," David Liu, the Harvard chemical biologist who led the work, told the *LA Times*, "32,000 of those are caused by the simple swap of one base pair for another."

Human germline engineering is another CRISPR application, which enables editing the DNA of an embryo itself—think designer babies. While germline engineering remains controversial—now think

Gattaca—it could mean ridding families of scourges like cystic fibrosis and sickle cell anemia, making it a medical advance as potentially important to this century as vaccines were to the last.

There's also stem cells to consider. One of the body's main repair mechanisms, stem cells have the remarkable ability to turn into any other type of cell, which is why the body uses them to repair worn-out tissue. Stem cell therapy works the same way.

Currently, there are only a handful of approved stem cell therapies in the United States, but this doesn't account for the incredible amount of work being done in labs all over the globe. Researchers are pioneering treatments for cancer, diabetes, arthritis, heart disease, macular degeneration, skeletal tissue repair, pain management, neurological diseases, auto-immune conditions, burns and other skin diseases, blindness, and much more.

What's more critical here is it's not just stem cells or gene therapy or CRISPR—it's the combined power of all these techniques, their convergence, that holds the most potential.

Perhaps the biggest consequence of this convergence will be individually customized medicine or what's called "N-of-1 medicine." In N-of-1 medicine, every treatment you receive has been specifically designed for you—your genome, transcriptome, proteome, microbiome, and all the rest. It's a level of preventative care not seen before. You'll know the foods, supplements, and exercise regimen that are perfect for you. You'll understand which microbes inhabit your gut, and what diet keeps them healthy and fit. You'll know which diseases you're most likely to develop and be able to take steps to prevent them. It's an era of incredibly personalized medical care, where the tools of life have become tools for the preservation of life, and many of the diseases that plagued earlier generations have begun to fade from memory.

CHAPTER FOUR

The Acceleration of Acceleration

The only constant is change, and the pace of change is accelerating—this is the argument we've been making. This increasing pace is the result of three overlapping amplifiers. First, the exponential growth of computing power and all the technologies explored in the last two chapters that are riding atop this growth. Next, accelerating individual technologies are converging with other accelerating technologies, producing overlapping waves of change that threaten to wash away almost everything in their path—or what happens when AI and robots converge and hundreds of millions of jobs disappear.

The final amplifier we want to explore is an additional set of forces, seven in total. Each of these is a by-product of converging exponentials, a "second order effect" in the technical parlance, that acts as an additional innovation accelerant. While each force functions independently, their real effect is combinatory. Think of them as steps in a mathematical equation, an algorithm designed (in a sense by all of us) to increase the rate of change in the world and the scale of its impact. Each step acts upon every other, each one amping up the next, all of them together accelerating our acceleration, producing more change in a year than our grandparents experienced in a lifetime.

Over the course of this chapter, we're going to explore these forces independently, while Part Two of this book will look at their impact collectively, seeing how these intersections will reshape our lives over the next decade. But, for now, let's take things one force at a time, starting, in fact, with time itself.

Force #1: Saved Time

In "The Original Macintosh," a collection of online anecdotes about the creation of that fabled machine, Apple computer scientist Andy Hertzfeld recounted a typical Steve Jobs story. It's typical, because, in this story, like in so many others, Jobs was frustrated.

The issue was speed.

The first Mac was supposed to be a very fast computer. And it was, at least on paper. Built around Motorola's 68000 microprocessor, the system was actually ten times faster than the Apple II. But it had limited RAM, making it necessary to upload extra info via floppy discs. And this was especially true during startup—which would occasionally drag on for minutes.

The delay was driving Jobs nuts. One day, he stormed into engineer Larry Kenyon's cubicle with a typical Jobsian demand: "The Macintosh boots too slowly," he said. "You've got to make it go faster."

Kenyon listened patiently. He'd heard it before. Once again, he told Jobs all the various ways they could make the computer speed up. Tinker with this, toggle with that. But, unfortunately, all of that tinker-toggling took place after startup.

Jobs wasn't pacified.

"You know," he continued, "I've been thinking. How many people are going to use this Macintosh? A million? No, more than that. In a few years, I bet five million. . . . Well, let's say you can shave ten seconds off of the boot time. Multiply that by five million users and that's fifty million seconds, every single day. Over a year, that's probably dozens of lifetimes. So if you make it boot ten seconds faster, you've saved over a dozen lives. That's really worth it, don't you think?"

Over the coming months, they did manage to shave ten seconds off of the boot time. And Jobs wasn't wrong: Those extra seconds did manage to save people time. But this wasn't an isolated incident. There's a pattern here, as "saved time" turns out to be one of the major benefits of technology.

Put differently, it's not just startup times that have shrunk.

Consider the search engine, one of our most widely used technologies. Prior to its arrival, if you wanted to know something, you went to a library—which took time. How much? Back in 2014, University of Michigan behavioral economist Yan Chen gave participants a bunch of questions, provided half of them with internet access and the other half with a library. Then she started the clock. Online inquiries averaged seven minutes an answer, offline required twenty-two, meaning every time we type a query into a search engine, technology saves us fifteen minutes. If we apply a little Jobsian logic to this, simply taking the 3.5 billion queries a day processed by Google, we find that Google alone saves us 52.4 billion minutes a day. That is to say, Jobs was right—that's a lot of lifetimes.

And the same can be said about the time saved via online shopping, entertainment, and all the rest. Buying a watch used to require, you know, going to the store. Watching a movie meant riding in a car to a theater. Booking a plane ticket once included telephone calls, hold times, and sometimes, flesh-and-blood humans. Not anymore, and this has consequences.

Innovation demands free time. A few centuries back, one major reason the world changed so slowly was we lacked the time to make a difference. Most of our day was spent on basic necessities: growing or catching food, carrying water, sewing, darning, scrubbing, etc. Yet, as Jobs pointed out, technology solves this problem.

Over the past hundred years, labor-saving devices—a term that once meant electricity, running water, and appliances—have managed to reduce housework, our agreed upon least favorite activity, from fifty-eight hours a week in 1900 to 1.5 in 2011. For entrepreneurs and inventors, this is like getting a week's worth of work back, for free, every month, simply by being alive.

What this actually means is that saved time isn't just a benefit of technology, it's also a driver of innovation—another force accelerating our acceleration. And the time we're saving today pales in comparison to what tomorrow will bring. In the late 1800s, New York to Chicago was four weeks by stagecoach. A few decades later, trains reduced that to roughly four days. Airplanes shrunk it to four hours. But a few years hence, the Hyperloop will be doing that trip in under an hour, and virtual reality and avatars have the potential to take that to zero.

Sensors add intelligence to our appliances, but they also add hours to our lives. Consider that soon, when you run out of coffee, your refrigerator will notice and order more. A blockchain smart contract will place that order and an Amazon drone will bring it to your house. The only time you notice you're out of coffee is when you're moving the bag from your outside delivery box to your kitchen cabinet. Of course, pretty soon your very own butler-bot will do the coffee moving for you.

Where the biggest advantages will start to accumulate is in our working lives. In fields ranging from material science to medical research, by allowing us to test new compounds inside a computer rather than a laboratory, AI collapses discovery times from years to weeks. Quantum computing does more of the same, only faster. 3-D printing removes months and months from product fabrication, building construction . . . well, you get the picture.

All of this has an impact on the rate of innovation. As this bonanza of extra hours continues to pile up, inventors, entrepreneurs, those proverbial gals and guys in the garage, will get far more time to experiment, fail, pivot, fail again, pivot again, and, eventually, get it right. Technology has shrunk innovation development times and expanded the time innovators can devote to development. It's an accelerating feedback loop of acceleration—but it's not the only one.

Force #2: Availability of Capital

It was one of the greatest one-two combo-punches in history. In 1957, the Soviets landed the first punch when they launched Sputnik 1 into orbit. Afterward, all hell broke loose. Edward Teller, father of the hydrogen bomb, called it "the greatest defeat for America since Pearl Harbor." Senator Mike Mansfield warned: "What is at stake is nothing less than our survival." But the Soviets went even further, following up that first punch with another, just four years later, when Yuri Gagarin became the first person to orbit the Earth. In America, these two blows caught everyone on the chin, deepening the chill of the Cold War and igniting the space race.

So how did we fight back? With cash. A lot of cash.

A few months later, President Kennedy responded by creating the Apollo program, pouring 2.2 percent of the US GDP into the aerospace industry. This influx of greenbacks supercharged an era of innovation that took only eight years to go from Alan Shepard's suborbital hop to Neil Armstrong's lunar footprints.

Of course it did.

Nothing accelerates technological development like money. More bucks, more Buck Rogers. More cash means more people experimenting, failing, and, eventually, driving breakthroughs. And this brings us to our next force: an unprecedented rise in the availability of capital.

Today, it is easier for innovators to find funding than ever before. And this abundance is financing even more innovation—more moonshots, more crazy ideas, just plain more. Money might not make the world go round, but it certainly makes the future faster. So where is all this cash coming from?

Digital technology.

While new technology has always meant new ways to make money, digital technology gave us a critical variation on this theme: new ways to *raise* money. Crowdfunding was the first of these and represents, in terms of dollars spent, the low end of our capital availability spectrum.

For those unfamiliar, crowdfunding is pretty straightforward. The "crowd" in question refers to the billions of people currently online. The funding part means asking that crowd for money. Typically, a crowdfunder presents their product or service to the world, usually via a video posted to a dedicated site like Kickstarter, and asks for money in one of four forms: as a loan (technically peer-to-peer lending), as an equity investment, in exchange for a reward (e.g., a T-shirt), or as an advanced purchase of the proposed product or service.

And it can add up to a lot of money.

The very first crowdfunding project took place in 1997, when the British prog-rock band Marillion raised $60,000 through online donations to finance a US tour. Twenty years later, the size of that market had grown considerably, reaching, by 2015, a worldwide total of $34 billion. And while Marillion had to invent the entire back-end process that drove their campaign, today's entrepreneurs can choose from any of the six hundred different crowdfunding platforms available in North America alone.

Kickstarter, for example, one of the most popular reward-based platforms, has launched over 450,000 projects, with over $4.4 billion pledged to the site. It has also sped up the startup process considerably. The most successful Kickstarter campaign to date, a smart watch called Pebble Time, raised just over $20 million in little over a month—something that would have taken years to accomplish in Marillion's day.

And like many other digital platforms, crowdfunding is riding atop Moore's Law and experiencing double-digit growth. By 2025, experts project the total amount of money moving through the ecosystem will rise to $300 billion. Yet the biggest development isn't in the amount of cash involved; rather, it's who gets access to that cash.

Peer-to-peer microlending sites like Kiva have brought available capital to parts of the world where investors have long turned a blind eye, while reward-based programs have given us everything from difficult to fund ocean cleanup technologies to pie-in-the-sky breakthroughs such as Oculus Rift. By democratizing access to capital, crowdfunding allows anyone, anywhere, with a good idea and access to a smartphone, to seek the cash they need to get going. It's why Goldman Sachs described crowdfunding as "potentially the most disruptive of all the new models of finance."

If crowdfunding is the new way for entrepreneurs to raise capital, then venture funding, our next category, represents the old. Yet the old has always played a large role in accelerating the new. Over the past five decades, we have venture capital to thank for Apple, Amazon, Google, and Uber, among many others, making it not just a force accelerating our acceleration, but one of the foundational drivers of the process.

In the United States, venture funding rose from $8.1 billion in 1995 to $61.4 billion in 2016. And then came the banner year of 2017. In the US, investments reached $99.5 billion (the second highest total in history, with the highest being $119 billion spent in 2000, during the dot-com boom). But the bigger story was the rest of the world. Asia, a relatively new player, peaked at $81 billion, while European VCs set an all-time high: $21 billion.

More importantly, large chunks of this money are flowing directly into technology, greasing the wheels of innovation even more. VC investment in serious technology is especially prevalent. Blockchain has seen serious increases in recent years, as have voice-activated interface technologies (like Alexa). AI is also on the rise, with investments climbing from $5.4 billion in 2017 to $9.3 billion in 2018. And biotechnology experienced a similar boom, rising from $11.8 billion in 2017 to $14.4 billion in 2018.

Yet, when it comes to raising staggering sums of money in an eyeblink, little can compare with initial coin offerings (or ICOs). Emerging out of the cryptocurrency realm, ICOs are a new form of crowdfunding underpinned by blockchain technology. Startups can raise capital by creating and selling their own virtual currency—called either "tokens" or "coins." These tokens give you ownership in the company (or, at least, voting power) and the promise of future profits, or can take the form of a security, representing fractional ownership of a piece of real estate or the like.

ICOs have become famous for raising money quickly, in very large amounts, and under some strange circumstances. Filecoin, for example, is a blockchain-based decentralized data storage network that allows participants to lend out the extra storage space on their servers in exchange for Filecoins (the name of their token). When it launched its ICO in August 2017, the project raised $257 million in only thirty days. The first $135 million was raised in the first hour alone. Yet they didn't even have a working product.

Far from an isolated incident, one month prior to Filecoin's success, Tezos, a self-governing currency (billed as a Bitcoin-update), raised $232 million in just thirteen days. Then there's the EOS token, one of the most popular cryptocurrencies trading today, which brought in a record-breaking $4 billion from its yearlong ICO.

And these token trends are not slowing down. The number of ICOs per quarter has also ballooned, from roughly a dozen during the first quarter of 2017, to over a hundred by the last quarter of 2017, and there's been even more activity since.

Yet, forget ICOs for a moment. When it comes to the mother lode of deployable capital, the real heavyweight title belongs to sovereign wealth funds (SWFs). These investment behemoths hold an estimated $8.5 *trillion* in assets. That's trillion, with a "T."

Traditionally, SWFs have invested money in public equities, infrastructure and natural resources, but as the economic promise of startups continues to climb, these funds are increasingly hunting outsized returns on entrepreneurial shores. In 2017, according to the Sovereign Wealth Lab research center at Madrid's IE Business School, forty-two SWF deals valued at around $16.2 billion flowed in this direction.

And this pales beside Softbank CEO Masayoshi Son's mega-fund, the "Vision Fund." Driven by his belief in the "Singularity"—Ray Kurzweil's idea that developments in AI will lead to unprecedented technological growth and unfathomable changes for civilization—Son decided to try to accelerate this process.

"I totally believe [in] this concept," he said in a recent speech. "In next thirty years, this will become a reality. I truly believe it's coming, that's why I'm in a hurry—to aggregate the cash, to invest."

And aggregate cash is what he did. The Vision Fund got started in September 2016, when Mohammed bin Salman, then the deputy crown prince of Saudi Arabia, flew to Tokyo in search of ways to diversify his country's oil-dominated investment portfolio. There, Son pitched his idea: build the largest fund in history and use it to finance technology startups. Less than an hour later, bin Salman agreed to become the cornerstone investor. "Forty-five minutes, $45 billion," Son later said on *The David Rubenstein Show*. "One billion dollars per minute."

Quickly thereafter, companies like Apple, Foxconn, and Qualcomm joined in. And that only brings us to today. According to Son, the $100 billion Vision Fund is just "the first step." He's already announced that he's working to establish a second Vision Fund in the next few years. "We will briskly expand the scale. Vision Funds 2, 3 and 4 will be established every two to three years. We are creating a mechanism to increase our funding ability from 10 trillion yen to 20 trillion yen to 100 trillion yen."

Any way you slice it, that's a lot of yen. Add it to the pile already created by crowdfunding, venture capital, and initial coin offerings, and you start to see this availability of capital as much more than just the thrum of business. It's a technology turboboost that's turning dollars and cents into ideas and innovation at a record-setting pace.

Force #3: Demonetization

In the last chapter, we introduced the Six Ds, or the developmental stages through which all exponential technologies pass. We introduced these stages as temporal markers, ways to keep track of where a tech is today and where it's heading tomorrow. Here, we want to return to one of those stages—demonetization—to explore how it's also acting as a force for acceleration.

Let's start with the simple fact that innovation demands research. So what's better than having millions of research dollars at your disposal? How about having those millions of dollars stretch a million times further.

This is what demonetization provides.

In 2001, as we learned last chapter, sequencing the entire human genome took nine months to complete and cost $100 million. Today, Ilumina's latest generation sequencer can do that in an hour and for $100—or 6,480 times faster and a million times cheaper. As a result, if you're working in genomics, then your government research grants now go much further than ever before, accelerating insights and catalyzing breakthroughs.

And what is true for gene sequencing is true in dozens of fields. Tools once accessible to only the wealthiest companies and the largest government labs are now available at near-zero prices to just about anyone. The obvious example is the supercomputer in your pocket. That would have been a multimillion-dollar machine a few decades back. In *Abundance*, we calculated the amount of technology—music players, video cameras, calculators, etc.—that came free of charge inside what was then a fairly expensive smartphone ($800 or so) at over a million 2012 dollars. Today, the average $50 smartphone found in Mumbai has all the same equipment. Of course it does. Sensors like cameras, accelerometers, and GPS have shrunk a thousandfold in size and a millionfold in price.

Not too long ago, robots were the sole province of big business. Today you can buy one to vacuum your home for less than you'd spend on a vacuum cleaner. We're also seeing the electricity needed to power those robots drop in price. Renewables, according to a 2019 report by the International Renewable Energy Agency, now account for one-third of the world's power, and their cost has dropped below coal. At the current rate of decline, solar is five doublings away from being able to produce enough power to meet all of our energy needs. Eighteen months later, after solar doubles again, we'll meet 200 percent of our energy needs with this tech alone. We're heading toward a total demonetization of the power that powers the planet. Since all innovators require power, in itself this will further accelerate the rate of change in the world.

But to really increase that rate of change, accelerated innovation alone won't get it done. Someone always needs to bring these innovations to market. And thanks to demonetization, nearly every basic business requirement—energy, education, manufacturing, transportation, communication, insurance, and labor—is growing exponentially cheaper. More bucks means more Buck Rogers. But demonetization provides much more bang for that buck—which is, after all, how you get to warp speed.

Force #4: More Genius

In 1913, Cambridge mathematician G. H. Hardy received an unusual letter in the mail. "Dear Sir," it began, "I beg to introduce myself as a clerk in the Accounts Department of the Port Trust Office at Madras on a salary of 20 pounds per annum." The letter went on to offer nine pages of mathematical ideas, including 120 different results in numbers theory, infinite series, continued fractions, and improper integrals. "Being poor," the letter concluded, "if you are convinced there is anything of value I would like to have my theorems published. . . ." It was signed S. Ramanujan.

Now, it isn't unusual for a Cambridge mathematician to receive equations in the mail, but this letter piqued Hardy's interest. Although the math started out in familiar calculus, it quickly went in startling directions, reaching conclusions that, as Hardy later remarked, "must be true because, if they were not true, no one would have the imagination to invent them."

Thus began one of the wildest tales in math. Born in Madras in 1887, Srinivasa Ramanujan's mother was a housewife, his father a clerk in a sari shop. While he showed an early aptitude for numbers, Srinivasa had no formal training or real access to teachers. He also had little patience for academia. In college, he failed every class but math, but even his math professors couldn't understand his work. He flunked out before turning twenty and spent the next four years in extreme poverty. Finally, in desperation, at age twenty-three, he wrote that letter to Hardy.

Confusion—that was Hardy's first reaction. He showed the letter to a colleague, the mathematician John Littlewood, trying to figure out if it was a joke. It didn't take long for them to figure out it wasn't a joke. The philosopher Bertrand Russell ran into the duo the next day, finding them, as he later wrote, "in a state of wild excitement because they believe they have found a second Newton, a Hindu clerk in Madras making 20 pounds a year."

Hardy brought Ramanujan to Cambridge. Five years later, he was elected to the Royal Society, making him both one of their youngest members in history and the first from India. Before dying four years later of tuberculosis, Ramanujan contributed over 3,900 formulae to math, including solutions to problems long considered unsolvable. He also made critical contributions to computer science, electrical engineering, and physics. Resoundingly, he's considered one of history's great minds, an unabashed genius. But of all his accomplishments, perhaps most amazing is the fact that he was noticed at all.

Until recently, most genius was squandered. Even if you were born with incredible talents, the chances of you being able to use those abilities were limited at best. Gender, class, and culture mattered. If you weren't born wealthy and male, your chances of getting more than a third-grade education were slim to none. And even if you did manage to get enough education to unlock your talent, being recognized for that talent, being able to use it to make a difference—as Ramanujan discovered—was no simple feat.

While IQ isn't the only metric for genius, the standard distribution of the Stanford-Binet scale shows that only 1 percent of the population qualifies. Technically, this makes for 75 million geniuses in the world. But how many of them actually get to make an impact?

Until recently, not that many.

One of the by-products of our hyper-connected world is that these extraordinary individuals will no longer be casualties of class, country, or culture. We don't tend to think about opportunity costs surrounding lost brilliance, but they're arguably substantial. Yet, thanks to our increasing interconnectivity and the exponential explosion of networks, all of these barriers to discovering genius are starting to fall. The results will be more breakthrough ideas, faster innovation, and greater acceleration.

And this is only half of the story.

While genius may be a rare phenomenon, we are starting to understand its underlying neurobiology. There are two major strains to this work, a near-term and a far-term approach. In the near term, research into what might be called "the neurological basis for innovation"—that is, creativity, learning, motivation, and the state of consciousness known as flow—have allowed us to amplify these critical skills like never before.

Consider the nine-dot problem, a classic test of creative problem-solving. Connect nine dots with four lines in ten minutes without lifting your pencil from the paper. Under normal circumstances, fewer than 5 percent of the population can pull this off. In a study run at the University of Sydney in Australia, none of their test subjects did. But then the researchers took a second group of subjects, and used transcranial direct stimulation to artificially mimic many of the changes produced during flow. What happened? Forty percent solved the problem—a record result.

The long-term approach takes a similar tack, using technology to improve cognitive function, only soon the technology will be permanently implanted in our brains. Entrepreneurs such as Elon Musk, who started a company called Neuralink, and Braintree cofounder Bryan Johnson, who founded Kernel, alongside established companies like Facebook, have been pouring hundreds of millions of dollars into next-generation brain implants. The implants have been dubbed "neuro-prosthetics" or "brain-computer interfaces," and the goal, as Johnson explains, "is not about AI versus human. Rather, it's about creating HI, or 'human intelligence,' the merger of humans and AI."

Everybody agrees that Cyborg Nation is still a long ways off, but progress is moving faster than many suspect. We already have brain-computer interfaces that can help stroke victims regain control over paralyzed limbs, and others that help quadriplegics use computers simply by thinking. Sensory replacement devices are already here (think cochlear implants), and full-scale visual prosthetics—that final horizon—are coming this decade.

Memory is the latest frontier. In 2017, USC neuroscientist Dong Song borrowed the seizure-control neural implants used by epilepsy patients. By repurposing them to stimulate the neuronal circuits involved in learning and retention, Song produced a 30 percent boost in memory. In the near term, this is a new treatment for Alzheimer's; in the long run, it's brain-enhancement for everyone.

Ray Kurzweil famously pegged the development of the full cyborg to the middle 2030s. Kurzweil's averaging an 86 percent success rate for his predictions, but even if he's off by a decade, with the progress we're already seeing in everything from networks to neuroscience, the end result is more genius, more breakthroughs, and more acceleration.

Force #5: Communications Abundance

Our next innovation accelerator is the power of the network—a tool that allows minds to connect with other minds, exchange ideas, and spark invention. In *Abundance*, we explored how the rise of the coffeehouse in eighteenth-century Europe became a critical driver of the Enlightenment. These egalitarian establishments drew people from all walks of life, allowing novel notions to meet and mingle and "have sex," as author Matt Ridley famously wrote. By becoming a hub for information sharing—a network—coffee shops were foundational in driving progress forward.

Not surprisingly, we see similar network effects in cities, which are essentially coffee shops writ large. Two-thirds of all growth takes place in urban environments because population density leads to the cross-pollination of ideas. This is why Santa Fe Institute physicist Geoffrey West discovered that doubling the size of a city produces a 15 percent increase in income, wealth, and innovation (as measured by the number of new patents).

But just as the coffeehouse pales in comparison to the city; so does the city pale in comparison to the globe. In 2010, roughly one-quarter of the Earth's population, some 1.8 billion people, were connected to the internet. By 2017, that penetration had reached 3.8 billion people, or about half the globe. And over the next half-dozen years, we are going to wire up the rest of humanity, adding 4.2 billion new minds to the global conversation. Soon, all eight billion of us, every single human, will be networked together at gigabit speeds.

If network size, density, and fluidity have turned cities into the best transformation engines we've yet managed to create, then the fact that we are about to link the entire globe into a single network means the whole planet is just a few years away from becoming the largest innovation lab in history.

Force #6: New Business Models

Traditionally, innovation means the discovery of breakthrough technologies or the creation of new products or services. But this definition doesn't capture some of the most potent innovation taking place today: the creation of new business models.

A business model is the systems and processes a company uses to generate value. For most of history, these models were remarkably stable, dominated by a few key ideas, upgraded by a few major variations on these themes. "The basic rules of the game for creating and capturing economic value were once fixed in place for years, even decades, as companies tried to execute the same business model better than competitors did," explains a 2015 article in the *McKinsey Quarterly*.

In the twentieth century, this added up to around one major business revolution per decade. The 1920s, for example, gave us "bait and hook" models, where customers are lured in with a low-cost initial product (the bait, say, a free razor) and then forced to buy endless refills (the hook, aka: razor blade refills). In the 1950s, it was the "franchise models" pioneered by McDonald's; in the 1960s we got "hypermarkets" like Walmart. But with the arrival of the internet in the 1990s, business model reinvention entered a period of radical growth.

In less than two decades, we've seen network effects birth new platforms in record time, bitcoin and blockchain undercut existing "trusted third party" financial models, and crowdfunding and ICOs upend the traditional ways capital is raised. What do all of these new models share? By significantly shortening the distance between "I've got a neat idea" and "I run a billion-dollar business," these models are more than an upgrade to our systems and processes, they're actually another force for acceleration.

More importantly, the scale of disruption is increasing. What began as accelerating and converging technologies has become accelerating and converging markets, meaning the business model changes of the last few decades are nowhere near the changes that are coming. But this doesn't mean we're blind to the future. We can now see seven emerging models that could end up defining business over the next few decades. Each is a revolutionary new way of creating value; each is a force for acceleration.

1. ***The Crowd Economy:*** This includes crowdsourcing, crowdfunding, ICOs, leveraged assets, and staff-on-demand—essentially, all the developments that leverage the billions of people already online and the billions coming online. All have revolutionized the way we do business. Just consider leveraged assets, which allow companies to scale at speed. Airbnb has become the largest "hotel chain" in the world, yet they don't own a single hotel room. They leverage (that is, rent out) the assets (spare bedrooms) of the crowd. These models also lean on staff-on-demand, which provides a company with the agility needed to adapt to a rapidly changing environment. Sure, this once meant call centers in India, but today it's everything from micro-task laborers behind Amazon's Mechanical Turk on the low end to Kaggle's data scientist-on-demand service on the high end.

2. ***The Free/Data Economy:*** This is the platform version of the "bait and hook" model, essentially baiting the customer with free access to a cool service (like Facebook) and then making money off the data gathered about that customer (also like Facebook). It also includes all the developments spurred by the big data revolution, which is allowing us to exploit micro-demographics like never before.

3. ***The Smartness Economy:*** In the late 1800s, if you wanted a good idea for a new business, all that was required was to take an existing tool, say a drill or a washboard, and add electricity to it—thus creating a power drill or a washing machine. In his excellent book *The Inevitable*, author and *Wired* cofounder Kevin Kelly points out that we're about to see an updated version of this economy, with AI replacing electricity. In other words, take any existing tool, and add a layer of smartness. So cell phones became smartphones and stereo speakers became smart speakers and cars become autonomous cars.

4. ***Closed-Loop Economies:*** In nature, nothing is ever wasted. The detritus of one species always becomes the foundation for the survival of another species. Human attempts to mimic these entirely waste-free systems have been dubbed "biomimicry" (if you're talking about designing a new kind of product) or "cradle-to-cradle" (if you're talking about designing a new kind of city) or, more simply, "closed-loop economies." A simple example is the company Plastic Bank, which allows anyone to pick up waste plastic and drop it off at a "plastic bank." The collector is then paid for the "trash" in anything from cash to WiFi time, while the plastic bank sorts the material and sells it to the appropriate recycler—thus closing an open loop in the life cycle of plastic.

5. ***Decentralized Autonomous Organizations:*** At the convergence of blockchain and AI sits a radically new kind of

company—one with no employees, no bosses, and nonstop production. A set of preprogrammed rules determines how the company operates, and computers do the rest. A fleet of autonomous taxis, for example, with a blockchain-backed smart contracts layer, could run itself 24-7, including driving to the repair shop for maintenance, without any human involvement.

6 ***Multiple World Models:*** We no longer live in only one place. We have real-world personae and online personae, and this delocalized existence is only going to expand. With the rise of augmented reality and virtual reality, we're introducing more layers to this equation. You'll have avatars for work and avatars for play, and all of these versions of ourselves are opportunities for new businesses. Consider, for example, the multimillion-dollar economy that sprung up around the very first virtual world, Second Life. People were paying other people to design digital clothes and digital houses for their digital avatars. Every time we add a new layer to the digital strata, we're also adding an entire economy built upon that layer, meaning we are now conducting our business in multiple worlds at once.

7 ***Transformation Economy:*** The Experience Economy was about the sharing of experiences—so Starbucks went from being a coffee franchise to a "third place," that is, neither home nor work, but a "third place" in which to live your life. Buying a cup of coffee became an experience, a caffeinated theme park of sorts. The next iteration of this idea is the Transformation Economy, where you're not just paying for an experience, you're paying to have your life transformed by this experience. Early versions of this can be seen in the rise of "transformational festivals" like Burning Man, or fitness companies like CrossFit, where the experience is generally bad (you work out in old warehouses), but the transformation is great (the person you become after three months of working out in those warehouses).

What all this tells us is that business as usual is becoming business unusual. And for existing companies, as Harvard's Clayton Christensen explains, this is no longer optional: "Most [organizations] think the key to growth is developing new technologies and products. But often this is not so. To unlock the next wave of growth, companies must embed these innovations in a disruptive new business model."

And for those of us on the outside of these disruptive models, our experience will be better, cheaper, faster. *Better* meaning new business models do what all business models do—solve problems for people in the real world *better* than anyone else. *Cheaper* is obvious. With demonetization running rampant, customers—and that means all of us—are expecting more for less. But the real shift is the final shift: *faster*. New business models are no longer forces for stability and security. To compete in today's accelerated climate, these models are designed for speed and agility. Most importantly, none of this is in any danger of slowing down.

Force #7: Longer Lives

Computers run our world and algorithms run computers, which begs the question: Where do algorithms come from? Fear of poetry, that's where. Fear of the maddening influence of poetry.

Ada Lovelace was born in 1815, in London, the daughter of the infamous genius poet, Lord Byron. When Byron deserted the family before Ada was a teenager, her mother took charge of her education. A very intelligent woman, Lady Byron hired tutors for her daughter, particularly emphasizing math and science skills—which was a fairly radical step at a time when women were not really allowed into either profession. But Ada's mother had an ulterior motive. She was absolutely convinced it was the arts, and specifically poetry, that had driven her husband, Lord Byron, insane. Lady Byron was worried the condition might be hereditary and thus steered her daughter away from anything that might bring on such troubles.

The tutoring paid off. In 1833, when Ada was seventeen, her life took a turn for the computational. It was then that she met Charles Babbage, who held the same chair at Cambridge University that had been once occupied by Isaac Newton and would later belong to Stephen Hawking. After learning of her love of math, Babbage invited Ada and her mother to see his "difference engine," his steam-powered calculating machine.

Stunned by what she saw, Ada was determined to understand it. She got a copy of the machine's blueprints from Babbage. And she studied. And studied. And when Babbage created the next version of the difference engine, this time renamed the "analytical engine," Ada was ready. The analytical engine was the concept for the world's first programmable computer, only this one powered by steam. Luigi Menabrea, an Italian engineer, had written a paper about Babbage's idea—in French. Ada decided to translate it into English and, at the urging of Babbage himself, add her own ideas to the mix.

Which is exactly what she did. But Ada's ideas included a new idea, a novel way for the analytical engine to perform calculations. Ada Lovelace wrote the world's first published computer program—our first algorithm. Unfortunately, maybe it was the strain of her discoveries, maybe it was just bad luck, but not long after she finished the translation, Ada fell ill. The world's first computer programmer and one of the most interesting minds in history was dead by age thirty-six.

And this raises a second question: How many of us die before we're done? What might Ada Lovelace or Albert Einstein or Steve Jobs have accomplished with an additional thirty years of healthy life? It's ironic that as we reach our later years, when we have the most knowledge, the sharpest skills, and the greatest number of fruitful relationships, old age takes us out of the game. This brings us to our final accelerating force, which is an attempt to solve this problem: the extended healthy human lifespan.

Extending the healthy human lifespan means increasing the number of years we're operating at peak capacity and capable of making the greatest contribution to society. It's the ability to chase our dreams for a much longer period of time. How much longer?

Two hundred thousand years ago, the average caveperson hit puberty around age thirteen, and had children not long after. By the time most of our ancestors were in their mid-twenties, their children were having children. Once that happened, since food was scarce and precious, the best thing you could do to ensure the survival of your lineage was not steal a meal from your grandkids. Thus evolution built in a fail-safe—a twenty-five-year (average) lifespan.

In the millennia that followed, little changed. By the Middle Ages, lifespan had crept up to thirty-one. By the end of the nineteenth century, we broke forty for the first time. It was at the turn of the twentieth that real acceleration emerged. Everything from the discovery of germ theory and the creation of antibiotics, to the implementation of better sanitation and the increased availability of clean water, radically improved childhood mortality. In 1900, 30 percent of all deaths in the United States were children under five. By 1999, it was 1.4. In parallel, the green revolution and better transportation networks increased average caloric intake, thus increasing lifespan yet again. The end result was a net gain of nearly thirty years, with average lifespan hitting seventy-six by the turn of the millennium.

Since then, our ability to recognize and treat the two largest killers—cardiac disease and cancer—has us living routinely into our eighties. And when we tackle neurodegenerative disease, research shows we can expect average lifespan to reach and perhaps exceed one hundred years. But many people believe we're not stopping there.

Convergence is fueling this conviction. The intersection of AI, cloud computing, quantum computing, sensors, massive data sets, biotechnology, and nanotechnology is producing a plethora of new healthcare tools. A dizzying array of entrepreneurial companies have started leveraging these tools to commercialize lifespan extension.

The first time most heard about these efforts was in September of 2013, when Google (now Alphabet) announced its latest startup, Calico. Headlines across the country proclaimed that the tech giant was squaring off against death itself. "Google's New Project to Solve Death" shouted *Time*; "Google Wants to Cheat Death" declared the

Atlantic. The real truth is a little more subtle. What Calico is actually doing is publishing papers with titles like "A window into extreme longevity; the circulating metabolomics signature of the naked mole-rat, a mammal that shows negligible senescence." But the real point is that more money and more minds than ever before—that is, Google-sized money and minds—are being spent on anti-aging.

And Google's got company.

We'll get into this in greater detail in a later chapter, but for now know that three main approaches are being investigated. First is "senolytic medicines." In order to prevent runaway cell division (i.e., cancer), the body normally halts this process after a set number of doublings. These shut-off cells, called "senescent cells," create inflammation, a significant cause of aging. That's why the Jeff Bezos–backed Unity Biotechnology is attempting to develop senolytic medicines that target and destroy these cells, restoring proper function to previously inflamed tissue. Of greater note, giving middle-aged mice these same medicines can extend their healthy lifespan by as much as 35 percent.

Next is the so-called "young blood" approach. Back in 2014, Stanford and Harvard researchers showed that blood infusions from young mice reversed cognitive impairments in old ones. Since then, a variety of companies have been trying to isolate and commercialize different components of this process. Elevian, for example, a Harvard University spinoff, is working on a young blood factor called GDF11. When injected into older mice, GDF11 has been able to regenerate their hearts, brains, muscles, lungs, and kidneys.

Stem cells, our third approach, are showing the most promise. Samumed, LLC, for example, is targeting the signaling pathways that regulate the self-renewal and differentiation of adult stem cells. If successful, their patented molecules should be able to regrow cartilage, heal tendons, remove wrinkles, and, by the way, stop cancer. This also explains why Samumed, a company still in stealth mode, has a $13 billion valuation.

A different approach is being pioneered by Celularity, a company founded by stem cell pioneer Bob Hariri (Peter is also a cofounder). Hariri's experiments demonstrate that, in animals, placental-derived stem cells can extend life 30 to 40 percent. The company's mission is to make this approach viable in humans, harnessing stem cells to amplify the body's ability to fight disease and heal itself.

What does all this add up to? Well, Ray Kurzweil often discusses the concept of "longevity escape velocity," or the point at which science can extend your life for more than a year for every year that you are alive. As far future as this sounds, according to Kurzweil, we are a lot closer than you might expect. "It's likely [that we're] just another ten to twelve years away from the point that the general public will hit longevity escape velocity."

We are creeping ever closer to a technological fountain of youth. Thus, the impact that each of us can make with a couple more decades is another force accelerating our acceleration. When coupled to our six prior forces, the resulting convergent combination is dizzying. We're heading toward a world of long-lived, AI-enhanced, globally interconnected humans—a world far different from the one in which we find ourselves today.

And this brings us to Part Two of this book. To help us make sense of the very different world ahead, over the next eight chapters, we're going to explore the impact converging exponentials and their secondary effects—that is, the seven forces discussed in this chapter—are having on the wider world. Of course, because there's no reasonable way to cover such a vast topic, we've focused our inquiry. The sectors of society covered in Part Two represent both the top ten contributors to our economy and those areas with the greatest impact on our daily lives.

Furthermore, to help keep things manageable, we've interspersed longer chapters with shorter ones. The next chapter is a lengthy examination of the future of shopping, for example, while the one following it is a shorter look at the adjacent field of advertising. In Part Three, we'll also explore energy and the environment. And when we're done, we'll have a 360-degree look at how converging exponentials are reshaping tomorrow.

PART TWO

THE REBIRTH OF EVERYTHING

CHAPTER FIVE

The Future of Shopping

The First Platform Play

It started with watches, it ended with hedge fund managers, and in between it reshaped a nation. What was it? The Sears catalog.

Richard Warren Sears was born on December 7, 1863, in Stewartville, Minnesota. His mother was a housewife, his father a failed gold miner turned successful blacksmith and wagon maker. But his newfound financial security didn't last. In Richard's early teenage years, his father lost all their money in a disastrous stock-farm investment. The family never recovered. A few years later, insult was added to injury and his father died suddenly, leaving Richard to support his sisters and mother. He was sixteen years old.

Out of desperation, Sears taught himself telegraphy, and found work with the railroad. While he rose to become a station agent, he still didn't earn enough money to support his family, which meant he was always looking for new ways to increase his income. One of those ways arrived in 1886.

A Chicago wholesaler had shipped a case of watches to a jeweler in Redwood Falls, Minnesota, but the jeweler had refused delivery. So the watches were still sitting at the railroad station where Sears worked.

Never one to miss an opportunity, Sears contacted the manufacturer with an offer to sell the watches himself. Their hard cost was $12 a piece, but stores typically sold them at double the price. Sears, though, lit upon an idea—charge less.

It was the birth of discount pricing, and Sears's timing was fortuitous. In the early 1870s, the nation was in the throes of the great railway expansion. For the first time in history, average citizens could travel hundreds of miles in a day. But keeping time during those journeys was an issue.

Back then, every city tracked time by its own clock, so America had three hundred different time zones. In 1883, wanting to ensure the trains ran on schedule, the railways decided to standardize matters. They divided the country into four time zones, and clocks were adjusted accordingly. Suddenly, especially for station agents, watches kept trains running on time. Everyone needed one. Plus, Sears's asking price? An incredible deal at $14.

Sears ended up making a $5,000 profit (over $120,000 today) on that first crate of watches. He was twenty-three years old and a newly minted entrepreneur. Using the profits from that initial order to purchase another crate, he founded the R. W. Sears Watch Company, advertised his wares in local newspapers, and set out to expand his territory.

In 1887, Sears hired a young watch repairman named Alvah Curtis Roebuck to assist him. A partnership bloomed, and the business blossomed. Four years later, the duo published their first catalog, a 52-page selection of watches and jewelry. Two years after that, the catalog was a 196-page wonder, especially for a country that had grown up rural. No longer selling just watches and jewelry, they expanded into everything from saddles to sewing machines, changing shopping forever.

Over the next decade, the company continued to grow, with a serious assist from the Postal Service. Many rural counties didn't have mail delivery, which had become a hot button issue. Congress felt that delivery costs would be prohibitive, and local retailers—the folks who ran the mom-and-pop stores that dominated the country—were fiercely protective of their natural monopoly.

But in 1896, Congress passed the Rural Free Delivery Act, opening up this new territory. Sears rushed in to fill the void, with the advent of the automobile, aiding the rush. A network of roads soon emerged, allowing mail-order merchants to reach nearly every home in America—which is what helped the Sears catalog become one of the most potent forces for democratization in US history.

Before these catalogs, local merchants played all kinds of games with both the availability and price of goods. The rich had access to high-quality merchandise, while the poor had to sift through the dregs. Sears rewrote those rules. His catalog didn't discriminate. Prices were clearly marked, and everyone—regardless of class, creed, or color—paid the same price.

This approach paid off. "Within 10 years," wrote journalist Derek Thompson in the *Atlantic*, "the book bloomed to more than 500 pages. Before long, the so-called Consumer's Bible read like an index of the entire U.S. economy. The company sold dolls and dresses, cocaine and tombstones, and even build-it-yourself houses. Decades before technology analysts started talking about platforms, Sears was the OG platform technology."

In 1915, at its height, the catalog was twelve hundred pages long, sold over a hundred thousand items, and generated over a hundred million a year in revenue. And this was just the beginning, since Sears went into retail next.

The 1920s brought a giant wave of urbanization. Americans moved from the country to the city, and Sears capitalized on the shift, opening over three hundred retail stores, in every major metropolis, before decade's end. By the middle of the next decade, one US dollar out of every hundred was spent at Sears. And there are important reasons for this incredible success.

Author Jeremy Rifkin points out that all major economic paradigm shifts share a common denominator. "At a moment in time," Rifkin told *Business Insider*, "three defining technologies emerge and *converge* [emphasis ours] to create . . . an infrastructure that fundamentally changes the way we manage power and move economic activity across the value chain. And those three technologies are new communications technologies to more efficiently manage economic activity, new sources of energy to power the economic activity, and new modes of mobility . . . to more efficiently move the economic activity."

This was the exact wave Sears rode to prominence. The US Postal Service was its communications technology, cheap Texas oil was its fuel source, and the automobile was its new mode of transport. Yet paradigm shift replaced paradigm shift, and we all know where that road ends. After a 132-year run, in the autumn of 2018, Sears filed for bankruptcy. Between 2013 and 2018, the company shuttered over a thousand stores, lost $6 billion in revenue, and left its hedge fund owners to sell what remained for spare parts.

So what happened?

Walmart happened. Even though Sears pioneered the idea of discounted goods, Walmart beat them at their own game. They out-demonetized and out-democratized Sears. They built their stores on cheaper land, paid cheaper wages, and sold cheaper goods. But most importantly, they saw the exponential writing on the wall.

"The tragic irony of the Sears saga," continues Thompson in the *Atlantic*, "is that communications technology, marshaled so brilliantly during Sears's rise, was instrumental in the company's downfall. In the 1980s, Walmart and other more modern retailers used digital technology to understand what shoppers were buying and to relay those findings to Walmart headquarters, which could place bulk orders for the new bestselling brands and products. With Walmart playing a superior game of sell-cheap-stuff-efficiently, Sears's fall was swift. In the early 1980s, it was five times as big as Walmart, by total revenue. By the early 1990s, Walmart was twice as big as Sears."

But this was only chapter two of retail disruption. History, of course, has a way of repeating itself. Amazon, a company founded about the same time Walmart was disrupting Sears, would blend the best of both models. They used the same postal service that Sears rode to prominence and the same communications tech that helped Walmart do the same. They became the everything store that replaced the everything store that replaced the everything store. And it's not just Sears and Walmart feeling the pinch.

To say that the retail space has changed over the past decade would be a stunning understatement. E-commerce giants such as Amazon and Alibaba are digitizing the industry, riding exponential growth to new heights. Meanwhile, many brick-and-mortar stores have joined Sears in bankruptcy—in 2017 alone, the total was 6,700. All one needs to do is look at the following table to understand the massive upheaval afoot.

COMPANY	2006 VALUE ($B)	2016 VALUE ($B)	2019 VALUE ($B)	06–18 % CHANGE
Sears	14.3	0.9	BK	–100%
JCPenney	14.4	2.75	.18	–98%
Nordstrom	8.5	10.2	4.8	–43%
Kohl's	18.8	9.4	7.3	–61%
Macy's	17.8	12.2	4.8	–73%
Best Buy	21.3	14.5	17.6	–17%
Target	38.2	42.1	54.4	+42%
Walmart	158.0	216.3	319.47	+202%
AMAZON	17.5	474.4	893.1	+5103%

And if this level of disruption looks unsettling, realize the e-commerce revolution is just getting under way. Even though online sales increased from $34 billion in Q1 of 2009 to $115 billion in Q3 of 2017, this growth spurt only accounted for 10 percent of total retail sales. Why? A great many people are still not connected to the internet.

In *Abundance*, we referred to the "Rising Billion" as those newly enabled digital masses coming online in the decade ahead, taking the connected measure of humanity from 3.8 billion in 2017 to 8.2 billion in 2025. The majority of these people won't frequent retail stores or shopping malls. Instead, for reasons that will become much clearer as we go along, they'll make their purchases digitally, via mobile devices, from the comfort of their homes. To put this in broader terms, because retail is nestled at the convergence of communications, energy, and transportation breakthroughs, it's a canary in a coal mine, ground zero for Rifkin's "next major economic paradigm shift." And one thing's for certain, shopping will never be the same.

AI and the Retail Experience

No one really understands the impact AI will have on retail, but once you start to get a handle on what's coming, it's fairly clear the unfair advantages it'll bring to store owners, splitting the market into two camps: those who make full use of AI, and those who declare bankruptcy.

AI makes retail cheaper, faster, and more efficient, touching everything from customer service to product delivery. It also redefines the shopping experience, making it frictionless and—once we allow AI to make purchases for us—ultimately invisible.

Let's begin with the basics: the act of turning desire into purchase. For most of us, this means going to the store and buying what we need. Some of us rely on online marketplaces, and sometimes those marketplaces deliver our desires, sometimes they don't. Now, if you're lucky enough to employ a personal assistant, you have the luxury of describing what you want to someone who knows you well enough to buy that *exact right thing* most of the time.

For most of us who don't, enter the digital assistant.

Right now, the four horsemen of the retail apocalypse are waging war for our wallets. Amazon's Alexa, Google Assistant, Apple's Siri, and Alibaba's Tmall Genie are going head-to-head in a battle to become the platform du jour for voice-activated, AI-assisted commerce. Note the glaring absence of any traditional retailers from this list. This is unlikely to change given the head start these horsemen have—already they've spent billions in the AI arms race.

We've seen a market shift like this before. Nokia was the world leader in cell phones, but when smartphones showed up, they ended up out of business. Why? They were in the phone business, but suddenly the phone business had become the computer business. With companies like Apple and Google as their new competitors, they could never catch up. And this brings us back to our predicted parade of retail bankruptcies.

For baby boomers who grew up watching Captain Kirk talk to the *Enterprise*'s computer on *Star Trek*, digital assistants seem a little like science fiction. But for millennials, it's just the next logical step in a world that is auto-magical. And as those millennials enter their consumer prime, revenue from products purchased via voice-driven commands are projected to leap from $2 billion today to $8 billion by 2023. And while we're still a ways away from making retail purchases completely frictionless, the data indicates the direction the trend is heading: On average, consumers using Amazon Echo spent more than standard Amazon Prime customers: $1,700 versus $1,300.

Perhaps there is no better example of the disruptive potential of digital assistants than the 2018 demo of Google Duplex. Every year, at the Google I/O conference, seven thousand attendees get together for three days of keynotes, code labs, and—the highlight—interactive demos. In Silicon Valley, product demos are where legends are born. In decades past, it was black turtleneck–clad Steve Jobs who *wait, wait, there's one more thing*'d his way into the history books. But in 2018, the soft-spoken Google CEO Sundar Pichai may have stolen back the crown.

Pacing the stage of Mountain View's Shoreline Amphitheatre, Pichai began his demo by pointing out that a big part of getting things done still involves making phone calls. "You may want to get an oil change," he noted, "maybe call a plumber in the middle of the week, or even schedule a haircut appointment. . . . We think AI can help with this problem."

Then, piped through the Shoreline's massive speakers, with follow-along captions projected on a giant screen behind him, Pichai played the audience a series of pre-recorded phone calls made by Google Duplex, their next generation digital assistant. The first call made a reservation at a restaurant, the second one booked a haircut appointment. The haircut appointment cracked everyone up, primarily because Duplex has natural language abilities which include dropping a long "hhmmmmm" into the middle of the conversation. In neither case did the person on the other end of the phone have any idea they were talking to an AI.

For certain, this issue of AI anonymity raised some red flags of the "If an AI can fool a receptionist, who else could an AI fool?" variety. In the days after the conference, the demo produced a backlash, and Duplex now announces itself as "Google's automated booking system." But the system's success speaks to how seamlessly AI can blend into our retail lives and how convenient it will continue to make them.

This is just the begining. The next retail arena being disrupted by AI is customer service. According to a recent Zendesk study, good customer service increases the possibility of a purchase by 42 percent, but bad customer service translates into a 52 percent chance of losing that sale forever—meaning more than half of us will stop shopping at a store due to a single disappointing customer service interaction. These are significant financial stakes. They're also problems perfectly suited for an AI solution.

The same technology Pichai demonstrated that can make phone calls for consumers can also answer phones for retailers—a development that's unfolding in two different ways. First, for organizations interested in keeping humans involved, there's Beyond Verbal, a Tel Aviv–based startup that has built an AI customer service coach. Simply by analyzing customer voice intonation, the system can tell whether the person on the phone is about to blow a gasket, is genuinely excited, or anything in between. Based on research conducted on more than seventy thousand subjects in more than thirty different languages, Beyond Verbal's app can detect four hundred different markers of human moods, attitudes, and personality traits.

Already it's been integrated in call centers to help human sales agents understand and react to customer emotions, making those calls more pleasant, but also more profitable. For example, by analyzing word choice and vocal style, Beyond Verbal's system can tell what kind of shopper the person on the line actually is. If they're an early adopter, the AI alerts the sales agent to offer them the latest and greatest. If they're more conservative, then it suggests items more tried-and-true.

Second, there are companies like New Zealand's Soul Machines working to replace human customer service agents altogether. Powered by IBM's Watson, Soul Machines builds lifelike customer service avatars designed for empathy, making them one of many helping to pioneer the field of emotionally intelligent computing. We'll explore this in depth a little later, but what's critical here is a single stat: 40 percent. With their technology, 40 percent of all customer service interactions are now resolved with a high degree of satisfaction and without any human intervention. And because the system is built using neural nets, it's continuously learning with every interaction—meaning that percentage will continue to improve.

The number of these interactions continues to grow as well. Software manufacturer Autodesk now includes a Soul Machine avatar named AVA (Autodesk Virtual Assistant) in all of their new offerings. She lives in a small window on the screen, ready to soothe tempers, troubleshoot problems, and forever banish those long tech support hold times. For Daimler Financial Services, they built an avatar named Sarah who helps customers with arguably three of modernity's most annoying tasks: financing, leasing, and insuring a car. Of course, in a future where "distributed autonomous organizations (DAOs)" will control fleets of autonomous taxis, there will soon come a point that a DAO's AI will be talking to Daimler's Sarah about that financing, leasing, and insuring. It's an AI-to-AI negotiation, no human required.

Yet, as we'll soon see, this isn't just about AI—it's about AI converging with additional exponentials. Add networks and sensors to the story and it raises the scale of disruption, upping the FQ—the frictionless quotient—in our frictionless shopping adventure.

Go, Go, Gone Are Cashiers

It's April 2026, a cold, rainy day in Chicago. You're supposed to meet your mother for lunch, but forgot your coat. On the Uber autonomous ride downtown, a quick online search reveals a shop selling those new eco-friendly vegan leather jackets you've heard so much about—the leather grown from stem cells and no cows harmed along the way.

You click the "interested" button on your screen, slide the phone back into your pocket, and forget about it. The store's AI interfaces with your phone's AI and automatically redirects your taxi. Upon arrival, you realize you're outside one of those throwback "craft retail" shops where they still employ actual humans. A woman named Sylvia meets you at the door, holding the vegan-leather coat you selected. The jacket fits perfectly—which isn't a surprise. A couple months ago you used the modified Wii sensor in your phone to map your body down to its every last flap and fold. Most shoes have weight sensors these days, so as your waistline fluctuates, the body map automatically adjusts. Long before you walked into that shop, both your phone and the store's computers knew your shape and size.

No need to wait in line to pay for the item either. A variety of cameras and sensors track both you and the jacket, so as you walk out the door, the price is instantly deducted from your bank account. Perhaps your cryptocurrency account. Plus, because those sensors also know that this is your first trip to this store, they try to tempt you into a second shopping adventure by texting you a digital coupon that will deduct 25 percent off your next purchase.

At the same time your transaction settles, sensors built into the rack where your jacket once hung alert the store's AI. The AI orders another jacket from the manufacturer and texts an employee to restock the now empty hanger. Also, turns out, this is the third vegan leather jacket sold in two days, so the inventory control system notices the trend and orders a couple of backup jackets in popular sizes just in case.

The above scenario isn't that far off. In fact, it requires little beyond the ever expanding impact of the Internet of Things (IoT) on our world, something that will happen almost automatically as more and more devices are connected to the internet. And it's quite an impact. By 2025, according to research done by McKinsey, the value of the IoT on retail will be somewhere between $410 billion and $1.2 trillion. Even better, most of this technology has arrived.

Automatic checkout, which frees customers from the drudgery of waiting, is already here. Amazon introduced Americans to it in January 2018, when the initial Amazon Go store opened for business in Seattle. The following year Amazon Go opened seven more stores and has plans for three thousand more by 2021. The *New York Times* describes passing through the store's turnstiles as "similar to entering the subway, with an [in-store] experience that is more closely akin to shoplifting."

Upon entering, visitors scan a QR code on their phone and the AI does the rest. Cameras track customer movement down the aisles and weight sensors built into the shelves do the same for the products. Just grab what you want, drop it into your backpack and head home. On your way out the door, the cost is automatically charged to your Amazon account.

Once again, this is about frictionless shopping. Long lines deter customers. Plus, cashiers cost money. By reducing staff requirements—the only employee in an Amazon Go store is the single human checking IDs near the liquor section—McKinsey estimates automated checkout will save retailers $150 to $380 billion a year by 2025. Which is why Amazon is not the only company chasing this cashier-less future; the San Francisco startup v7labs is helping all retail stores make this same transition, while Alibaba's cashier-less Hema stores were tested in China a full two years ahead of Amazon.

Smart shelf technology is already here, employing RFID (radio frequency identification) tags and weight sensors to detect when an item has been removed. The innovation deters theft, automates restocking, and ensures that inventory is always in the right spot. Today, Intel's version has a screen built into the shelf. Tomorrow, smart shelves will be AI-enhanced and capable of conversation. Is the sweater you're holding dry clean only? Just ask the shelf.

Perhaps the biggest shift in retail will be one of efficiency, especially in supply chain management. Back in 2015, a Cisco study found that IoT solutions will have more than a $1.9 trillion impact on this sector, and for good reason. AI can detect patterns in data that humans cannot. This means that every link in the supply chain—inventory levels, supplier quality, demand forecasting, production planning, transportation management, and more—is being revolutionized.

And fast.

Seventy percent of retail and manufacturing companies currently digitize every aspect of their logistics operations. More importantly, all this disruption is occurring even before the robots arrive in retail. But . . .

The Robots Are Coming, The Robots Are Coming

On August 3, 2016, the prayers of stoned Dungeons & Dragons players were answered. That day, Domino's Pizza introduced the Domino's Robotic Unit, or DRU for short. The first home delivery pizza robot, the DRU looks like a cross between R2-D2 and an oversized microwave. LIDAR and GPS sensors help it navigate, while temperature sensors keep hot food hot and cold food cold. Already, it's been rolled out in ten countries, including New Zealand, France, and Germany, but its August 2016 debut was critical—as it was the first time we'd seen robotic home delivery.

It won't be the last.

A dozen or so different delivery bots are currently entering the market. Starship Technologies, for example, a startup created by Skype founders Janus Friis and Ahti Heinla, has a general-purpose home delivery robot. Right now, the system is an array of cameras and GPS sensors, but soon models will include microphones, speakers, and the ability—via AI-driven natural language processing—to communicate with customers. Since 2016, Starship has carried out fifty thousand deliveries in over one hundred cities in twenty countries.

Along similar lines, Nuro, the company cofounded by Jiajun Zhu, one of the engineers who helped Google develop their self-driving car, has a miniature self-driving car of their own. Half the size of a sedan, the Nuro looks like a toaster on wheels, except with a mission. This toaster has been designed to carry cargo—about twelve bags of groceries (version 2.0 will carry twenty)—which it's been doing for select Kroger stores since 2018 (in 2019, Domino's also partnered with Nuro).

As these delivery bots take over our streets, others are streaking across the sky. Back in 2016, Amazon was first, announcing Prime Air, their promise of drone delivery in thirty minutes or less. Almost immediately, companies ranging from 7-Eleven and Walmart to Google and Alibaba jumped on the bandwagon. While critics remain doubtful, the head of the FAA's drone-integration department recently said that drone deliveries may be "a lot closer than . . . the skeptics think. [Companies are] getting ready for full-blown operations. We're processing their applications. I would like to move as quickly as I can."

While delivery bots are starting to spare us that trip to the store, for those who prefer shopping the old-fashioned way—i.e., in person—once you actually arrive at that store, robots will be there to help. Actually, they've been there for a while.

In 2010, SoftBank introduced Pepper, a humanoid robot capable of understanding human emotion. Pepper's cute: four feet tall, with a white plastic body, two black eyes, a dark slash of a mouth, and a base shaped like a mermaid's tail. Across her chest is a touch screen to aid in communication. There's been a lot of communication. Pepper's cuteness is intentional, as it matches its mission: help humans enjoy life as much as possible. Over twelve thousand Peppers have been sold. She serves ice cream in Japan, greets eaters at a Pizza Hut in Singapore, and dances with customers at a Palo Alto electronics store. More importantly, Pepper's got company.

Walmart uses shelf-stocking robots for inventory control, Best Buy uses a robo-cashier to allow select locations to operate 24-7, and Lowe's Home Improvement employs the LoweBot—a giant iPad on wheels—to help customers find the items they need while tracking inventory along the way.

The biggest benefit robotics provides might be in warehouse logistics. In 2012, when Amazon dished out $775 million for Kiva Systems, few could predict that just six years later there would be forty-five thousand Kiva robots deployed at all of their fulfillment centers, helping process a ridiculous 306 items *per second* during the Christmas season.

Many other retailers are following suit. Order jeans from the Gap and soon they'll be sorted, packed, and shipped with the help of a Kindred robot. Remember the old arcade game where you picked up teddy bears with a giant claw? That's Kindred, only her claw picks up T-shirts, pants, and the like, placing them in designated drop-off zones that resemble tiny mailboxes (for further sorting or shipping). The big deal here is democratization. Kindred's robot is cheap and easy to deploy, allowing smaller companies to compete with giants like Amazon.

For retailers interested in staying in business, there doesn't appear to be much choice in the matter. By 2024, the US minimum wage is projected to be $15 an hour (the House of Representatives already passed the bill, but the wage hike is meant to unfold gradually between now and 2025), and many feel that number is way too low. Yet, as human labor costs continue to climb, robots won't just be coming, they'll be here, there, and everywhere. It's going to become increasingly difficult for store owners to justify human workers who call in sick, show up late, and can easily get injured. Robots work 24-7. They never take a day off, never need a bathroom break, health insurance, or maternity leave. Going forward, this means technological unemployment will become more of an issue—and more on this in Part Three—but in retail, the robotic benefits for both companies and customers are considerable.

3-D Printing and Retail

In 2010, Kevin Rustagi was frustrated. So were his friends Aman Advano, Kit Hickey, and Gihan Amarasiriwardena. These four recent MIT grads had been thrust into the working world, where they quickly discovered they hated their outfits. Business clothes sucked. And it didn't make sense—athletes got to perform in all sorts of high-tech gear, but accountants had to make do with Dockers?

So they decided to bring a little whiz-bang to the boardroom. Together, they formed the Ministry of Supply, a clothing company intent on borrowing space suit technology from NASA for a line of dress shirts. In 2011, with an original fundraising goal of $30,000, a runaway Kickstarter campaign netted them almost half-a-million. They were in business.

Soon afterward, the "Apollo" dress shirt hit the market. Looking like a traditional button down, the Apollo is anything but. The shirt uses "phase change materials" to control body heat and reduce perspiration and odor. It also adapts to the wearer's shape, and stays tucked in and wrinkle-free all day. As *TechCrunch* summarized: "In essence, it's a magic shirt."

That magic shirt led to magic pants, suits, and more. Ministry of Supply now makes high-performance smart clothing for both sexes, including a new line of intelligent jackets that respond to voice commands and learn to automatically heat to your desired temperature. Recently, they extended their high-tech approach to manufacturing. Head into their Boston-based retail outlet on fashionable Newbury Street and you can have your high-performance shirt—or suit, blouse, or pants—3-D printed while you wait. It takes about ninety minutes. And the machine is a marvel. With four thousand individual needles and a dozen different yarns, the printer can create any combination of materials and colors desired, with zero waste.

And if you can't make it to Boston, not a problem. These days, if you want 3-D-printed clothing, all you need is a smartphone. Since fashion designer Danit Peleg's 2015 introduction of the first line of 3-D-printed clothing available via the Web, a half-dozen designers have followed suit. Both Reebok and New Balance deploy the technology, the former to upgrade the speed and quality of its manufacturing facilities, the latter to build custom insoles for athletes. Many other fashion houses are not far behind.

And fashion is only part of the story, as 3-D printing is now showing up all over retail. Staples, the office supply company, has been offering the service for years. They recently launched an online version where customers upload designs for office products from home, Staples employees print them in-store, and the final product is delivered to your door. The French hardware manufacturer Leroy Merlin has taken this a step further, allowing customers to print bespoke hardware in their stores. Need a ten-inch flat-head nail or a curving socket wrench made to reach around corners? They've got you covered.

And this is only where we are today. Over the next ten years, 3-D printing will reshape retail in four key ways:

1. ***The End of the Supply Chain:*** With 3-D printing, retailers can now purchase raw materials and print inventory themselves, either at warehouses or in the retail outlet. This means the end of suppliers, manufacturers, and distributors.

2. ***The End of Waste:*** Okay, maybe not the complete end of waste, but with consumers preferring eco-friendly products and retailers looking to minimize materials cost, the exactitude of 3-D printing is a ready-made solution.

3. ***The End of the Spare Parts Market:*** If you're a farmer in Iowa and your tractor breaks during harvest time, waiting a few days for a spare part could jeopardize the entire season. A 3-D printer solves this problem. And it'll solve the same problem for everything from coffeemakers to skateboard wheels. This doesn't just mean an end to the spare parts business, it also means a new level of longevity for the products we purchase.

4. ***The Rise of User-Designed Products:*** Sure, there will always be some version of Apple in the market—an uber design-centric company pushing out products so slick they always find a buyer. Yet, for everything from fashion to furniture, customer-designed will replace designer-designed as standard operating procedure.

But this does raise a final question: With Alexa placing our orders, 3-D printers manufacturing those orders, and drones delivering the results to your doorstep, why, in the not too distant future, would anyone go anywhere to go shopping?

Retail's Last Hope

In "Welcome to the Experience Economy," an article for the *Harvard Business Review,* author Joseph Pine tracks two hundred years of economic development via a curious metric: the birthday cake.

> As a vestige of the agrarian economy, mothers made birthday cakes from scratch, mixing farm commodities (flour, sugar, butter, and eggs) that together cost mere dimes. As the goods-based industrial economy advanced, moms paid a dollar or two to Betty Crocker for premixed ingredients. Later, when the service economy took hold, busy parents ordered cakes from the bakery or grocery store, which, at $10 or $15, cost ten times as much as the packaged ingredients. Now, in the time-starved 1990s, parents neither make the birthday cake nor even throw the party. Instead, they spend $100 or more to "outsource" the entire event to Chuck E. Cheese's, the Discovery Zone, the Mining Company, or some other business that stages a memorable event for kids—and often throws in the cake for free. Welcome to the emerging *experience economy.*

By replacing premade ingredients with premade experiences, the experience economy is a new kind of disruptive business model satisfying a new kind of need. For most of history, we didn't want prepackaged experiences because life itself was the experience. Just staying safe, warm, and fed was adventure enough. Technology changed that equation.

At the turn of the Industrial Revolution, the richest people on the planet didn't have air-conditioning, running water, or indoor plumbing. They lacked automobiles, refrigerators, and telephones. Plus, computers. Today, even folks living below the US poverty line draw on these conveniences. Those better off draw on much more. So much, in fact, that we've started to take our stuff for granted. As a result, for many, experiences—tactile, memorable, and real—have become more valuable than possessions.

Retailers have capitalized on this trend. Starbucks made bank extending the familiarity of the local coffee shop to a global scale. Outdoor retailer Cabela's turned their showrooms into faux outdoor adventures complete with waterfalls. And now converging exponentials will take the experience economy to new heights.

Consider the Westfield shopping center group's ten-year vision for the future of retail, "Destination 2028." Replete with hanging sensory gardens, smart changing rooms, and mindfulness workshops, Westfield's proposed shopping center will be a "hyper-connected micro-city" with an incredible amount of personalization. Smart bathrooms will provide individually customized nutrition and hydration tips; eye scanners and AI can personalize shopping "fast lanes" based on prior purchases; and magic mirrors will offer virtual reflections of you wearing an entire range of new products.

Combining entertainment, wellness, learning, and personalized product-matching, Westfield's "Destination 2028" aims to help make you into a better you—and they're betting that this is worth the inconvenience of leaving the house to do your shopping.

It's a big bet. In the US, there are over eleven hundred malls and forty thousand shopping centers. Minnesota's Mall of America is a small town, spanning 5.6 million square feet and housing five hundred stores. China's biggest mall covers over 7 million square feet and is larger than the Pentagon. This upgraded experience economy might mean these malls have a chance of staying in business.

But it'll be a very different business. If successful, retail will become a convergent industry, where time spent at the mall pays multiple dividends. Shopping becomes healthcare becomes entertainment becomes education and so forth. Or, as we'll see in the next section, our malls become a memory, as shopping itself becomes another task outsourced to your AI.

No More Shopping Malls

Earlier in this chapter we ran through a thought experiment about the year 2026, when sensors, networks, and AI will have converged to remake shopping. Here we want to run another experiment, pushing the clock forward a few more years and adding five additional technologies to the retail mix.

Welcome to April 21, 2029, a sunny day in Dallas. You've got a fundraising luncheon tomorrow, but nothing to wear. The last thing you want to do is spend the day at the mall. No sweat. Your body-image data is still current, as you were scanned only a week ago. Put on your VR headset and have a conversation with your AI, which you've conveniently named JARVIS (because your coauthor can't seem to shake his *Iron Man* fetish).

"It's time to buy a dress for tomorrow's event" is all you have to say.

In a moment, you're teleported to a virtual clothing store. Zero travel time. No freeway traffic, parking hassles, or angry hordes wielding baby strollers. Instead, you've entered your own personal clothing store. Everything is in your exact size. And we mean everything. The store has access to nearly every designer and design on the planet. Ask JARVIS to show you what's hot in Shanghai, and presto, instant fashion show. Every model strutting the runway looks like you, only dressed in Shanghai's latest.

The phone rings. It's your best friend. She joins you in the shop via her own VR headset. As you two chat, your AI listens. Your comments become commands: "I'd love some black pumps to match my new dress" makes racks of perfectly fitting shoes appear.

Yet none of these shoes seem exactly right. "How would this dress look with those satin Jimmy Choos sitting in your closet?" asks your friend. No problem. Every piece of physical clothing you own in the real world has a digital twin available in the virtual. You ask, and instantly, you're wearing them.

When you're done selecting your outfit, the AI pays the bill. While your new clothes are being 3-D printed at a warehouse—before speeding your way via drone delivery—a digital version has been added to your personal inventory for use at future virtual events. And the cost? With no middlemen, less than half of what you paid in stores.

Back in present day, let's break this future apart.

3-D body-scanning is already here. By using infrared depth sensing and imaging technology to produce an exact digital copy of the surface of your body, companies like Levi's and Bloomingdales have imaging booths in select stores, while brands like Nike, Boss, and Armani are not far behind. Nor is it just the big brands getting involved. Bombfell is a fashion subscription service for casual menswear that pairs human fashion experts with AI, selecting from over eighty different brands and sending them your way. Online retailers are also in the mix, with Amazon acquiring the 3-D body-scanning startup Body Labs in 2017 as a way to make bespoke clothing just another feature available through Prime Wardrobe.

As far as an AI fashion advisor goes, those too are here, courtesy of both Alibaba and Amazon. During their annual Singles' Day shopping festival, Alibaba's FashionAI concept store uses deep learning to make suggestions based on advice from human fashion experts and store inventory, driving a significant portion of the day's $25 billion in sales. Similarly, Amazon's shopping algorithm makes personalized clothing recommendations based on user preferences and social media behavior.

And the VR system itself? Well, right now there's Hololux, a collaboration between Microsoft and the London College of Fashion. Their VR goggles let you shop in mixed reality anywhere in the world. Want to check out the Prada store on London's High Street—once again, no problem.

So, there you have it, a future in which shopping is dematerialized, demonetized, democratized, and delocalized—otherwise known as "the end of malls." Of course, if you wait a few years after that, you'll be able to take an autonomous flying taxi to Westfield's Destination 2028—which might be an experience worth having, so, maybe, it's not the end of malls. Either way, it's a top-to-bottom transformation of the retail world.

Now, if we could only do the same thing to advertising . . . .

CHAPTER SIX

The Future of Advertising

Madder Men

In the Emmy Award–winning TV show *Mad Men*, a classic 1960s-era advertising agency was the center of the action. It was an agency filled with big egos, boozy lunches, and emerging technology, an era when print, television, and radio ads ruled the roost. And those Mad Men had a good run. For almost half a century, this trilogy of media defined how companies pushed their products to the public, and those agencies rode that wave until it crashed onto those game-changing shores known as the internet.

Certainly, when the dot-com revolution first arrived, few understood the disruption it would bring to advertising. Yet, in nearly no time at all, Craigslist gutted newspapers' classified sections, and banner ads—of all things—gutted magazines. Next came the spread of DVRs and paid digital services of the Hulu, Netflix, Amazon variety, a series of innovations that collectively saved humanity from the drudgery of TV commercials. And now, less than two decades after the arrival of the internet, Google and Facebook together command more advertising dollars than all print media on the planet.

In 2017, Google's ad campaign revenue totaled over $95 billion;

Facebook's reached over $39 billion. Taken together, this is roughly 25 percent of all global advertising expenditure. Fueled by open source e-commerce platforms, mobile devices, and advances in online payment infrastructure, social media marketing has replaced virtually the entire traditional advertising industry. That took fewer than fifteen years.

And the numbers are huge. In 2018, the global advertising industry surpassed $550 billion, driving Google's valuation north of $700 billion and Facebook's above $500 billion. The reason for this spike is that these numbers are based on our data and the telling trail left by our searches: our likes and dislikes, what we desire, who our friends are, and what we (and they) are clicking on these days.

But with a blitzkrieg of technologies converging on the industry, advertising will continue to change. First, it's likely to get a little more invasive and a lot more personal. Yet this won't last. Not long after, the entire social media marketing market will vanish. How long will that take? We give it ten to twelve years.

Let's take look at why.

The Spatial Web

For all of human history, the experience of looking at the world was roughly the same for everyone. Other than mental illness, psychedelic drugs, or an overactive imagination, reality was a shared constant—what you saw out there was what I saw out there. But now the boundary between the digital and the physical is beginning to fade. The world around us is gaining layers of information. Invisible without the right apparatus, but don a pair of AR glasses and you'll find rich, personalized, and interactive data where none has ever been before. And this means my world and your world are now very different worlds.

Welcome to Reality 2.0, or Web 3.0, or the Spatial Web. To understand the Spatial Web, it helps to start with the first Web, version 1.0, where static documents and read-only interactions meant that the best way advertisers could reach consumers was through banner ads. Web 2.0 was an upgrade, introducing multimedia content, interactive Web ads, and participatory social media. Yet all this was still mediated by 2D screens. Web 3.0 is the next phase. Because of the convergence of high bandwidth 5G connections, augmented reality eyewear, our emerging trillion-sensors economy, and, to stitch it all together, powerful AI, we have gained the ability to superimpose digital information atop physical environments—freeing advertising from the tyranny of the screen.

Imagine stepping into a future Apple Store. When you approach the iPhone display, a full-sized AR avatar of Steve Jobs materializes. He wants to give you a tour of the product's latest features. Avatar Jobs is a little too much, so with nothing more than a voice command, he's replaced with floating text—and a list of phone features hovers in the air in front of you. After you've made your selection, eschewing the iPhone for a new pair of AR iGlasses, another voice command is all it takes to execute a smart contract.

Next, glasses on, you head over to a friend's house. While chatting in her kitchen, you gaze at her new cabinets. Sensors in the glasses track eye motion, so your AI knows your focus has been lingering. Via your search history, it also knows you've been considering remodeling your kitchen. Because your smart recommendation preferences are turned on, cabinet prices, design, and color choices fill your field of vision. It's a new form of advertising, either an extension of frictionless shopping, or a novel type of spam, depending on your perspective.

The early version of this reality is already here. Known as "visual search," the feature is currently available from an assortment of companies. A partnership between Snapchat and Amazon, for example, allows you to point their app-camera at an object, then get a link showing either the product itself or something similar, available for purchase. Pinterest, meanwhile, has a multitude of visual search tools, such as Shop the Look, which dots every object in a photo. Like the couch? Click the dot. The site will find you similar products for sale. Or take Lens, their real-time visual search tool. Point the app-camera at a scene and the app will generate links to all the products in that scene.

Google takes this one step further. Released in 2017, their Google Lens is a general visual search engine. It does more than just identify products for sale, it decodes an entire landscape. You can learn anything you want: the botanical breakdown of the plants in a flowerbed, the breeds of dogs romping through a park, the history of the buildings lining a city street.

IKEA has taken things the farthest. By using their AR app via smartphone, you can map your living room. This gives you a digital version complete with all the furniture in its exact dimensions. Need a new coffee table? Their technology lets you try out different styles in different sizes. Your choice triggers a smart payment, and just like that, IKEA customizes your coffee table and delivers it to your doorstep. Need help assembling? Their AR app can walk you through it step-by-step.

All this visual search competition has kicked development into overdrive, spiking consumer adoption rates as well. As more people use these systems, more data is fed back to the AI running them. By fall of 2018, this feedback loop had pushed visual searches above a billion queries a month. Pretty much every global brand is preparing for a world of "point, shoot, and shop." It's another reason we may be looking at the end of the shopping mall, because reality becomes the shopping mall. And if you think this sounds invasive—well, wait, wait, wait, there's downright creepy to consider.

The Eerie Power of Hyper-Personalization

You've been spotted. You're just out for a casual stroll through a department store, and their facial recognition system has you in its sights. Your AR glasses light up: "Hi, Sarah, great to see you. . . ."

Damn, you forgot to change your preferences to "Do Not Disturb."

A microsecond later, the store's TV monitors continue the assault. Maybe it's a hologram of the President of the United States calling out your name, "Sarah, just one second. Your pores are a matter of national security. I want to tell you that your genome sequence matches a new line of L'Oréal skincare products."

When you don't respond to POTUS, the AI switches tactics. Now it's Mom. You flinch, involuntarily. Her voice is deeply imprinted on your brain. But you know better, and just keep walking. Heck, sometimes it's your favorite movie stars (based on data from your Netflix account), or your favorite sports star (based on internet searches). The one that freaks you out the most? Father McFarland, your priest. Whatever the case, it's a far cry from the Marlboro Man. If it weren't so annoying, it might actually be funny.

Does this sound like a far-off fantasy? Guess again.

Sure That Sounds Like Mom, But Can You Prove It?

Remember *Mission: Impossible*? Remember the throat mic that Tom Cruise used to mimic the voice of the bad guy? Well, impossible no more. Two companies have demonstrated real-life versions. The Montreal-based startup Lyrebird—named after the sound-imitating bird—has a novel speech synthesis technology which allows them to mimic the voice of someone with very little data.

Thirty sentences is all it takes. A three-minute video at your surprise birthday party—hell, you don't even remember anyone filming—is more than enough. And researchers at the Chinese search giant Baidu have an AI that works faster than Lyrebird. Just ten 3.7-second samples is enough to train a voice-imitation system more than 95 percent of the time. A hundred five-second clips and the match will be almost perfect.

While Lyrebird and Baidu's speech synthesis technologies are not yet completely convincing, the field is rapidly advancing—and not just to make advertising creepier than normal.

"There are plenty of great use cases for the technology," says Leo Zou, a member of Baidu's communications team, in *Digital Trends*. "Voice cloning could help patients who lost their voices. It's also an important breakthrough for personalized human-machine interfaces—a mom can easily configure an audiobook reader with her own voice, for example. [Additionally,] the method allows for original digital content. Hundreds of characters in video games will have unique voices because of this. Or speech-to-speech language translation, as the synthesizer can learn to mimic a speaker in another language."

Deeper Fakes

In 2018, a YouTube video of former President Barack Obama made its way around the internet. Over 6 million people tuned in to see him seated beside an American flag, speaking earnestly to the camera. "President Trump," he explained, "is a total and complete dipshit. Now, I would never say these things. At least not in a public address. But, someone else would, someone like Jordan Peele."

Then the video shifts to a split screen. On the left, Obama continues talking. On the right, we see actor-director-comedian Jordan Peele actually speaking the words being put into the former President's mouth. The video is a deepfake, an AI-driven, human image synthesis technique that takes existing images and videos—say, Obama speaking—and maps them onto source images and video—such as Jordan Peele imitating President Obama insulting President Trump.

Peele created the video to illustrate the dangers of deepfakes. He felt the need to make it because it's really just one of thousands. Political chicanery, revenge porn, celebrity revenge porn—they've all been tried and tried again. And while these early versions are pretty close to the genuine article, their fakery is detectable.

Researchers at Carnegie Mellon University have recently developed new algorithms capable of far greater realism. Their AI not only transfers head position, facial expression, and eye gaze from one video to another, it transmutes subtle details—eye blink rate, slight eyebrow flicks, the micro-wiggle of a shoulder. And with fewer glitchy distortions. The results are convincing. Study subjects overwhelmingly thought the videos were real.

While there are some positive uses for the technology (which we'll examine in the upcoming entertainment chapter), the application's downside is impossible to ignore. Many are concerned about "fake news," which has the power to destroy reputations, set off civil unrest, and even shape global politics. There are also legal ramifications. Deepfakes make it easier for those "caught on tape" to claim it wasn't them—because you never know—and that's exactly the problem.

Yet, the problem in advertising—that is, being followed around department stores by marketers disguised as Mom—is more a temporary blip than a longstanding trend. Pretty soon advertising itself might actually go away.

Goodbye Advertising, Hello JARVIS

From the Mad Men of old to the Madder Men of today, the purpose of advertising hasn't changed: to sell you stuff. So ads extol benefits: Buy X because it'll make you Y—sexy, successful, shiny, whatever. But what happens when you are no longer the one making the buying decisions? Oh yeah baby, it's Shopping JARVIS to the rescue.

Imagine a future when you simply say: "Hey JARVIS, buy me some toothpaste." Does JARVIS watch TV? Did he happen to catch those late-night ads filled with bright-white smiles? Of course not. In a nanosecond, JARVIS considers the molecular formulations of all available options, their cost, the research that supports their teeth-whitening claims, the research that condemns their breath-freshening claims, published client-satisfaction reports, and finally—maybe later rather than sooner—it evaluates your genome to determine the flavor formulation most likely to tingle your taste buds.

Then it makes a purchase.

Taking it a step further, in the future, you'll never actually have to order toothpaste. JARVIS will be monitoring your supply of regularly consumed items—from coffee, tea, and almond milk to toothpaste, deodorant, and all the rest—and will order supplies before you realize what needs restocking.

How about purchasing something new? That drone your son wants for his birthday? Just specify functionality. "Hey JARVIS, could you buy me a drone for under a hundred dollars that is easy to fly and takes great photos?"

What about fashion decisions? Will we trust our AIs to choose our clothes? Seems unlikely, until you consider that AIs can track eye movement as we window-shop, listen to our daily conversations to understand likes and dislikes, and scan our social feeds to understand our fashion preferences as well as those of our friends. With that level of detail, Fashion JARVIS will do a damn good job of selecting our clothing—no advertising required.

We're heading toward a future where AI will make the majority of our buying decisions, continually surprising us with products or services we didn't even know we wanted. Or, if surprise isn't your thing, just turn that feature off and opt for boring and staid. Either way, it's a shift that threatens traditional advertisers, while offering considerable benefits to the consumer.

CHAPTER SEVEN

The Future of Entertainment

Going Digital

One way to tell the story of the rise of digital entertainment is through four key decisions made by Reed Hastings, founder and CEO of Netflix. The first of these was made in 1999, when Hastings was a computer scientist turned entrepreneur. He'd already taken his first software company public, then taken a hefty buy-out, leaving him a sizable nest egg to plunge into his next company. Hastings lit upon an interesting idea: Rent DVDs over the Internet and use the postal service to deliver them. He decided to give it a try—which was the decision that birthed Netflix.

The second decision came a few months later, when Hastings had a bigger idea: Never charge late fees. The third decision was Netflix's real breakthrough, the killer app, the "queue." Subscribers could create a list of movies they wanted to watch and as soon as the company received word that a prior DVD had been mailed back, Netflix would send out another. Because their DVD rental policy allowed three movies to be checked out at once, users were never without something to watch, and the convenience turned the company into everyone's default movie rental house. It was one of the earliest platform plays, and it helped transform the industry, making Netflix a titan.

In 1999, the year Netflix launched, they had 239,000 subscribers; only four years later they hit a million. But it was Hasting's fourth decision—his 2007 choice to replace the postal service with broadband streaming—that was the real game-changer. By fall 2018, subscriptions had rocketed to 137 million, and experts predict that could double over the next few years.

Netflix is now the eight-hundred-pound streaming gorilla. Fifty-one percent of all streaming subscriptions flow their way, earning them over $4.5 billion in annual revenue and $150 billion in market cap. But it's what they're doing with all that money that's even more disruptive.

Netflix is creating content. A lot of content. In 2017, they spent $6.2 billion on original movies and TV shows, outspending major studios such as CBS ($4 billion) or HBO ($2.5 billion), and just shy of the $8–10 billion-a-year range of heavyweight contenders like TimeWarner and Fox. A year later, Netflix doubled their spend to $13 billion, placing them in the ranks of the majors.

But, again, the real news is what they did with that money. In 2018, while the big six movie studios released a combined seventy-five films, Netflix's war chest produced eighty new features and over seven hundred new TV shows.

This is why any understanding of the impact of exponentials on entertainment should start with Netflix. In fact, in the annals of exponential disruption, the company's assault on Blockbuster is now a classic tale. The fact that Blockbuster passed up the opportunity to buy Netflix for $50 million is perhaps only matched by Kodak's failure to capitalize on digital photography, the very technology they'd invented. Yet, Netflix's assault on Blockbuster was merely the result of a *solitary* convergence.

To kill home video, Netflix leveraged a new network, the internet, to allow Americans to order DVDs from the comfort of their couch. Today, Netflix is utilizing a short stack of converging exponentials—particularly broadband and artificial intelligence—to take on the trillion-dollar entertainment ecosystem.

And it's not just Netflix.

Streaming platforms are exploding. Most big technology companies are getting in on this action, the byproduct of converging technologies leading to converging markets. In 2018, Apple spent over $1 billion on original programming, while Amazon dropped $5 billion. Sling, YouTube, Hulu, even that guy who repairs lawnmowers and has three million followers on Facebook—they're all coming to eat Hollywood's lunch.

Throughout the rest of the chapter, we'll explore how converging exponentials will remake entertainment over the next decade. Three major shifts—call them who, what and where—are underway. We're seeing alterations in *who* is making content, *what* kind of content is being made, and *where* we're experiencing that content.

Since the inception of the silver screen, entertainment has primarily been the product of a few well-capitalized, tightly controlled studios and networks. The combination of TV ad sales and box office revenue produce just shy of $300 billion a year. By hoarding a few scarce resources—tech, talent, financing, and distribution—a handful of Hollywood studios and TV networks have maintained a virtual stranglehold on those dollars.

But accelerating exponentials have a way of making scarce resources abundant. This time is no different. And this brings us to the first of three major shifts in entertainment, a shift in *who* makes content.

The Rise of the Uber-Creator

In the early 2000s, as videos cameras, editing systems, and audio recorders started to become standard on our phones, people began doing something both obvious yet unexpected—using these tools to make content. Continents of content. We coined a new phrase for these continents: user-generated content. Blogs captured the written side of this exchange; podcasts, the audio form. However, video was an issue. There was no one place these user-generated videos could call home, a hub from which everything could be freely shared.

Companies raced to fill that gap, Google foremost among them. The tech giant was desperate to be first, but their video sharing service was stuck in legal hell. Lawyers were freaked about rights. What was Google supposed to do if users posted content they didn't own?

YouTube, though, didn't have that problem. Back then, the company was just three ex–Pay Pal employees with an idea, a garage, and a credit card. They were just too small to care about lawyers.

They didn't stay that way. While Google dithered about who could post what, YouTube exploded. Less than six months after YouTube cofounder Jawed Karim posted "Me at the Zoo," the rather unremarkable first video to go up on the site, a clip of Brazilian soccer phenom Ronaldinho became the first to garner a million views. This led to a $3.5 million investment from Sequoia Capital, which YouTube used to upgrade their network and cement their position. A little more than a year later, Google decided it was easier to join forces than compete. They shut down their video sharing service and shelled out $1.65 billion to purchase YouTube, calling the site "the next step in the evolution of the Internet."

Understatement.

Every day, billions of people watch billion of videos on the site. For the younger generation, YouTube has completely replaced television as their medium of choice. Meanwhile Hollywood's longtime stranglehold on talent has been shattered as YouTube democratized content distribution for all of us. Social media influencers are the result, a new breed of uber-creators challenging traditional media in untraditional ways.

Take cooking shows. Celebrity chefs like Gordon Ramsay and Rachael Ray are now up against YouTube shows like *Binging with Babish*, where host Andrew Rea recreates meals from famous TV shows and movies for over a million viewers an episode. Or *Cooking with Dog*, where the kitchen escapades of a silent Japanese woman are narrated by her poodle, Francis, for the pleasure of millions. Or the 800,000 who regularly tune in to *My Drunk Kitchen*, which is pretty much as it sounds.

And these new stars are bringing in big bucks. In 2018, YouTuber Logan Paul racked up $14.5 million for his comedic vlogs, while gamer Daniel Middleton (DanTDM) took home $18.5 million. Neither is working alone. Musicians are also making bank, as are kids playing with toys. A seven-year-old named Ryan, the star of *Ryan ToysReview*, takes in over $22 million a year, earning him the number one spot on *Forbes*'s list of highest paid YouTube entrepreneurs. Plus, since nothing makes money like money, the venture crowd has gotten in on the action. Upfront Ventures, Khosla Ventures, First Round Capital, Lowercase Capital, SV Angel, and more, are all betting on user-generated content—meaning the uber-creator has now become as bankable as big name Hollywood stars.

As technologies continue to converge, the scale of disruption will only increase. The smartphone video camera was an uprising, allowing the masses to become makers. Then platforms like YouTube gave those makers a playground, and a way to get paid. But there are now app-based services like Bambuser that help anyone make their own live-streaming broadcast network, a development that allows creators to take aim at entire entertainment ecosystems.

Blockchain will amplify this process. By allowing artists to create unchangeable digital records of their work (making piracy impossible), and because its transaction costs are negligibly low or nonexistent, blockchain is bringing us to that fabled land of content creation: micropayments. This is what writers, artists, filmmakers, comics, and journalists have been waiting for since the internet first arrived. Direct to fan, no middlemen. A true meritocracy of creativity—or so the story goes.

Right now, new content platforms are popping up to capture all this energy. Niche markets are everywhere. Pretty much anything you can watch or listen to other people do—coding software, building robots, petting cats—there's a channel for that, on-demand or livestreamed, and backed up with apps that allow new levels of fan interactivity. The most unusual development isn't just the uber-empowerment of creators, it's also the kinds of creators being empowered.

In June of 2016, the extremely eerie short film *Sunspring* was released online, the end result of a neural net–powered AI being fed hundreds of sci-fi film scripts and allowed to take a crack at writing one of its own. Two months later, Twentieth Century Fox debuted the trailer for the upcoming thriller *Morgan*, also created with the help of an AI—this time, IBM's Watson.

To pull this off, Watson "watched" trailers for a hundred horror movies, then conducted visual, audio, and composition analysis to understand what humans deemed scary. By applying this same kind of analysis to *Morgan*, the AI identified the film's critical moments. Although a human was needed to arrange those moments into a coherent order, Watson reduced the amount of time it takes to make a trailer from ten days to one.

Movies aren't the only format entering the machine age. Researchers at the Georgia Institute of Technology have developed Scheherazade, an AI that creates Choose Your Own Adventure–style stories for video games. While current AI-driven video games start out with a fixed number of datasets, thus a fixed number of possible storylines, Scheherazade enables unlimited plot points. It's literally an infinite adventure machine—only, it's not all about the algorithms. Scheherazade has help, human help. Content creation is done via collaboration between an AI and the crowd.

Which brings us to the second shift in entertainment, a shift in the kind of content being created.

From Passive to Active

The next major shift in entertainment is in *what* kind of content is being made. Over the next three sections, we'll see that content is about to become much more collaborative, immersive, and personalized. We'll break each of these down in turn, but the place to start is with the death of "passive" media.

Passive media means information flows in only one direction. It's traditional newspapers, magazines, television, movies, and this book. Active is the opposite. It means the information flows both ways, and finally, the user gets to have their say.

Active media isn't new. Many companies now treat users as developers. The Wikipedia page for "Video games with user-generated gameplay content" lists ninety-five separate titles and is definitely incomplete. Popular titles such as Doom and Mario Maker include easy-to-use map editors so anyone can construct their own levels and share them online. But AI-gaming technologies like Scheherazade take this interactivity to whole new heights.

Also, into other media.

Enter MashUp Machine, an AI-driven platform for participatory storytelling. By blending machine intelligence with crowd intelligence, this app creates interactive animated movies. And it's a two-way street. As users customize content, the AI learns the ins-and-outs of their storytelling styles, allowing it to make suggestions and help the process along.

Quality will only continue to improve. With machines helping us tell stories, our machines will become better storytellers. Soon, AI won't just scan content for relevant topics and sticky memes to produce more content. Instead, it'll digest entire novels, consume entire films, and—by inputting enough storytelling—know how to distinguish diamonds from dung.

At the same time, we humans might be losing that skill. Deepfakes are an obvious example. What began as a disturbing trend in politics and pornography has spread to other forms of entertainment. It's an entirely new kind of collaborative, active media.

In 2018, researchers at the University of California, Berkeley, developed an AI motion transfer technique that superimposes the bodies of professional dancers onto the bodies of amateurs, lending their fluid movements to your normal cha-cha-cha. This means anyone can become Fred Astaire, Ginger Rogers, or Missy Elliott. It's a full-body deepfake, with a key difference: democratization.

Deepfake version 1.0 used AI-driven frame-by-frame image transfers that required multiple sensors and cameras. To pull off these dance fakes, all you need is your smartphone's camera.

These fakes bring a host of opportunities to entertainment—like raising the dead. How long before Hollywood studios start reanimating Robin Williams, Marilyn Monroe, or Tupac Shakur? How long before we get new movies with old stars? Our hunch, not very long.

And then there's the real fakes side of the deepfakes discussion: the use of computers to create alternative versions of ourselves. We already have AI-driven personal assistants: Siri, Echo, and Cortana. Imagine, you just got in a spat with your significant other and could really use some advice. Yet saying, "Hey, Siri, my boyfriend is mad at me," only produces "I don't know how to respond to that." But what if your personal digital assistant were the renowned life coach Tony Robbins.

You don't have to imagine. In 2018, Robbins teamed up with Lifekind, a company that specializes in creating AI "personas" of real people. These are audio- and photo-realistic simulations indistinguishable in everything from manners to memory. To recreate Robbins, Lifekind blended over 8 million images, with the complete library of his work—books, videos, blogs, podcasts, and tapes of live events. The result, according to Robbins, is "an operating AI, not a bot. It won't be able to do therapy [yet], but I'll be able to make that happen [eventually]. . . . Already, the audio's so good that my wife can't tell it's not me. But the most interesting part is the actual AI. The opportunity to capture how a person thinks, feels, or creates is extraordinary. It has a memory capacity that dwarfs anything I have. And it has all my [self-help] models, so it can look at you . . . and using [a particular] model decide that 20 percent of you is concerned, 30 percent is excited, 40 percent is engaged. It can literally do this in real time." Not only is content becoming more active than ever before, those activities are blending human and machine intelligence to expand the entertainment industry into wild new terrain.

The Holodeck Is Here

Jules Urbach went to high school with Rod Roddenberry, the son of Gene Roddenberry, the creator of *Star Trek.* The two became close friends. They talked almost every day. What did they talk about? "The holodeck," says Urbach. "A lot of the time, we talked about the holodeck."

Introduced in *Star Trek: The Next Generation*, the holodeck uses holograms to produce almost any experience a user desires. It's a fully immersive environment, functionally indistinguishable from real life. Urbach's obsession became a mission—build a holodeck in the real world.

That mission took him into video games, then 3-D gaming, and finally 3-D rendering. Urbach cofounded Otoy, a company that figured out how to move rendering off the desktop and into the cloud. Before they came along, an effects-heavy film like *Planet of the Apes* required hours of supercomputer processing to produce a single frame of film. With Otoy's software, this happens in real time, on a tablet phone, connected to the cloud via WiFi.

Next, Urbach cofounded LightStage, a company that specializes in photo-realistic 360-degree image capture. This technology is also what turns people into holograms, providing Otoy with the basic images needed for special effects.

But there are still two hurdles to clear before a holodeck can become reality. The biggest one is light. When we see an object, we're seeing trillions of photons bouncing off that object. So if you can artificially project trillions of photons toward the eye, at just the right angle and intensity, you can recreate reality, any reality.

Enter Light Field Lab, a California-based startup manufacturing the first-ever display technology able to generate those trillions of photons. While their initial displays are only four- by six-inches across, they're capable of projecting a holographic image two inches thick, viewable from an area thirty degrees wide. By combining these cubes into eighteen-inch displays and then combining those displays into wall panels, Light Field Lab can fill an entire room with these cubes: walls, floor, and ceiling. And every one of them can project a hologram ten feet out. It's the *Star Trek* holodeck.

Or almost. Since objects in the holodeck feel like real objects, the final hurdle is touch. And here too, Light Field has made progress. In the same way that they use light to make you see, they use sound to make you feel. When ultrasound—yes, the same technology used by doctors—is projected into the room, sound waves themselves can give objects a physical presence. It's not quite the heft of real objects, but it's tangible. And by combining Otoy's software, LightStage's image capture, and Light Field's projector, we have all the basic components for a holodeck—the most immersive form of entertainment yet.

Immersion marks the third shift in content, a transformation that has everything to do with attention. When it comes to attention, active trumps passive, and immersive trumps active. The reason is sensory input. The more senses engaged in an activity, the more attention we pay to that activity.

That's why companies are devising all sorts of ways to drag our senses into the virtual. Haptic gloves that engage our sense of touch are already here, and increasingly sophisticated. There are also scent-emitting devices to bring Smell-O-Vision into television, and 3-D audio systems that provide concert-style experiences from the comfort of our living rooms. Haptic chairs now pitch and yaw and shiver and shake, while *Ready Player One*–style omnidirectional treadmills allow us to dance, dance, dance in any direction we want.

Stanford neuroscientist David Eagleman wants to take this further. He's teamed up with Second Life creator Philip Rosedale, to expand haptic sensation from the hands to the entire torso. Rosedale's latest creation, High Fidelity, is a fully-immersive VR world. Eagleman's startup, NeoSensory, has built an "exoskin" that's designed to work in that world. It's a long-sleeved shirt with micro-motors placed every few inches along the arms, back, and stomach. "If it's raining in the VR world," explains Eagleman, "you can feel the raindrops. Or if you're touched by another avatar, you can feel their touch." And the signaling is so fast that the wearer feels that touch instantaneously.

Further still, the Los Angeles–based Dreamscape combines haptic-sensing with immersive VR to allow whole groups to share exotic encounters—such as swimming through the deep ocean beside blue whales or petting creatures in the Alien Zoo. And since Dreamscape has partnered with AMC theaters, we're not long from the day when participatory moviegoing replaces big-screen blockbusters as the summer's best thrill ride.

Urbach's holodeck is the next giant leap forward. The use of ultrasound to provide a sense of touch will provide this same level of haptic sensing, but without the need for fancy gloves. The AI powering that holodeck will also be emotionally aware, and the environment it will create will be incredibly interactive, meaning all three of our major shifts in entertainment will be bundled into one experience. This will mark a sizable change in entertainment, but it's not the end of this story.

Things are also about to get personal, really, really personal.

This Time It's Personal

It's 2028, the end of a long day, and no rest for the wicked. You have less than forty-five minutes to get ready for dinner, but first you want to sit down, have a drink, and be entertained. Do you pick up the remote and flick through channels? Doubtful. Are you interested in a holographic CNN floating above the coffee table? Nope. But the good news is that neither of these questions matter—because your AI already knows what you need.

Not only has your AI been with you all day, it now has the ability to monitor and understand your emotions. It's followed the highs and lows of your mood in considerable detail. It caught your grimace in a smart mirror early this morning, heard the angry conversation you had with your wife around lunch, and was there on the ride home when you ignored a call from your brother. This last bit is especially telling, because your AI has been around long enough to know you only ignore calls from your brother when really stressed. Plus, sensors have been tracking your neurophysiology along the way, so not only does the system understand the details of your emotional life, it knows how those details impact your body and brain.

And it can act on all this information.

The moment you walk into the living room, projected onto the walls are your favorite clips from Owen Wilson comedies. What makes this unusual is you didn't actually know you were an Owen Wilson fan, but over the past five years, you've caught a handful of his old movies, often without realizing it. And while the movies might have been less than memorable, each one had that scene that cracked you up. Your AI noticed. It also knows, because it's been tracking your emotional history, that laughter—in 78.56 percent of all prior high-stress situations—was your fast track to feeling better.

So now you're treated to a parade of Owen Wilson scenes, plus a few from other movies with the same style of comedy. Near the end of the session, your AI also inserts a few videos from your phone—movies of you and your wife laughing together. Happy memories that remind you of what's actually important. And the mash-up works perfectly. By the time you've finished your drink, that bad mood has lifted. You talk through the argument with your wife and are off to dinner feeling more energized than you've felt all day.

The crazy part: Most of this technology is already here, found under the label of "affective computing," or the science of teaching machines to understand and simulate human emotion. It's a tale of convergence, a new field sitting at the intersection of cognitive psychology, computer science, and neurophysiology, combined with accelerating technologies like AI, robotics, and sensors. Already, affective computing is seeping into e-learning, where AIs adjust the presentation style if the learner gets bored; robotic caregiving, where it improves the quality of robo-nursing; and social monitoring, like a car that engages additional safety measures should the driver become angry. But its biggest impact is on entertainment, where things are getting personal.

Facial expressions, hand gestures, eye gaze, vocal tone, head movement, speech frequency and duration are all signals thick with emotional information. By coupling next generation sensors with deep learning techniques, we can read these signals and employ them to analyze a user's mood. And the basic technology is here.

Affectiva, a startup created by Rosalind Picard, the head of MIT's Affective Computing Group, is an emotional recognition platform used by both the gaming and the marketing industry. The technology tells a customer service chatbot if a user's confused or frustrated, provides advertisers with a way to test the emotional effectiveness of their ads, and gives gaming companies a means to adjust play in real time. In the thriller game *Nevermind*, Affectiva's tech hunts anxiety via facial expressions and biofeedback. When the system discovers the user is scared, the game doubles down—adding challenging tasks and surreal content to up the thrill factor.

Lightwave, another emotional-computing startup, can capture not just the emotional state of an individual, but that of a whole crowd. It's already been utilized by Cisco to judge a startup pitch competition, helped DJ Paul Oakenfold increase listener engagement at a concert in Singapore, and measured viewer reactions during a pre-screening of *The Revenant*.

Affective computing is also going mobile, which means our phones are starting to serve up content based on what's going on in the real world—our mood, location, companionship, the mood of those companions, and on and on. Startups like Ubimo and Cluep are providing everything an emo-entrepreneur needs, from affective app development platforms to highly personalizable emotional-content delivery services.

As convergence continues, a new assortment of personalized possibilities arrive—not just content that's been presorted to fit our mood, but content that's been individually created to fit that mood.

The same AI-driven Choose Your Own Adventure–style storytelling invading video games has started making its way into traditional media. In May of 2018, 20th Century Fox announced they were turning the 1980s Choose Your Own Adventure book series into a big screen event. Throughout the movie, the audience used their smartphones to vote on which direction they wanted the film to go, choosing storylines, plot twists, and more. Unfortunately, having to make smartphone choices was an engagement killer—the *Hollywood Reporter* called it a "major disruption to the movie theater experience in the worst way possible." But that smartphone interface is only a temporary solution. Soon, with in-theater sensing and affective computing, emotionally driven storytelling will become just another part of the moviegoing experience.

Our AI will also know our plot preferences better than we know them ourselves. Consider that you may remember enjoying a movie, but your AI knows why you enjoyed it. With semantic analysis and biofeedback, it understands how a line of seemingly innocuous dialogue became a deeply powerful nostalgic moment. It knows your heart rate, blinking rate, pupil dilation response, where you're looking, where you're not. It's this kind of data deluge that means that even in choose-your-own-adventure, before long, we won't be the ones choosing. Our AIs will match our mood to our history, neurophysiology, location, social preferences, and desired level of immersion and then, in an instant, customize content to match them all.

And this brings us to our final shift, not a change in *who* is making the content or *what* kind of content is being made, but rather a revolution in *where* we'll be experiencing that content.

Here, There, and Everywhere

The story of story: a brief overview. Historians believe that storytelling itself began around the campfire, but they are certain that storytelling to the masses began with print. Books, newspapers, magazines were our first big-league carriers of information, earning them a four-hundred-year run at the center of our entertainment world. Radio came next, offering a level of intimacy and immediacy never seen before. Silent films and talkies were powerful additions, but radio was the very first technology that allowed an entire nation to tune in together.

Then black-and-white TV created an era of instantly shared visuals. Next came color television, which was less upgrade than conquest. These massive boxes earned a half-century monopoly on the center of our living rooms. Then plasma screens. At consumer electronics show after consumer electronics show, those screens got thinner, cheaper, and higher resolution. Next, cords got cut and those screens went everywhere, and what could possibly disrupt this disruption?

Enter the augmented reality of Magic Leap and the company's stated purpose: to eliminate the screen altogether. Sure, their first generation AR glasses were a piece of hardware so geeky we can only conclude they had unintentional prophylactic effects. But things have gotten sexier since, as have reasons for wearing those glasses. Magic Leap wants to dematerialize the screen, putting it anywhere you want it to be—the wall of your bedroom, the palm of your hand, the side of the Brooklyn Bridge.

So what could possibly disrupt that? How about an AR smart contact lens. No longer would you have to wear headgear, since the screen is mounted on your cornea and projects onto the back of your retina. And what could possibly disrupt that? How about the holodeck—no eye gear required. And what could possibly disrupt that? Well, as Steve Jobs already told us, wait, wait, there's one more thing. . . .

Let's take a closer look:

First, for those who still want screens, the technology itself is morphing. OLEDs (organic light emitting diodes) have begun replacing LEDs. The initial draw was image resolution, but the ultimate advantage is flexibility. LG already has a nineteen-inch OLED display that can be rolled into a cylinder, and other companies aren't far behind. We've seen something similar in phones, where swapping rigid silicon for malleable graphene has allowed Chinese researchers to develop a smartphone that can be wrapped around your wrist and worn as a bracelet. Added to this flexibility, we're seeing touch screens that provide tactile feedback. By flowing ultra-low-level electric current back into your skin, our touch screens can now touch back.

But screens have an inherent limitation: place. Screens mean watching entertainment in a fixed location—your living room or your local movie theater. Certainly, we get mobility via our tablets and smartphones, but the trade-off is size and, by extension, engagement. If we're watching content built for screens the size of billboards on phones the size of postage stamps, then we're much more likely to get sucked out of the story by distraction. With augmented reality, though, we're beginning to transition away from screens altogether.

This transition is coming quickly. Over the next five years, AR is projected to create a $90 billion market. "I regard [AR] as a big idea like the smartphone," Apple CEO Tim Cook recently said in an interview with the *Independent*. "The smartphone is for everyone. We don't have to think the iPhone is about a certain demographic, or country, or vertical market. It's for everyone. I think AR is that big, it's huge."

And what this growth buys us is an information layer projected atop regular reality. The world becomes the screen. If you want to play AR *Star Wars*, you're battling the Empire on your way to work, in your cubicle, cafeteria, bathroom, and beyond.

We got our first taste of this in 2016, when Nintendo released Pokémon GO and the greatest cartoon character turkey shoot in history ensued. With 5 million daily users, 65 million monthly users, and over $2 billion in revenue, the popularity of that experience remains one for the record books. In the years since, apps have exploded. Glasses that were once thick and bulky have become thin and light. They're also about to become a whole lot smaller. Companies like the well-funded startup Mojo Vision are developing AR contact lenses that will give you heads-up display capabilities and more, no glasses required.

And for those who have no interest in AR contact lenses, Urbach projects that the first versions of his holodeck will start to show up in Disney-style theme parks and possibly in rec rooms for the extraordinarily wealthy by decade's end. But who needs a holodeck when you can start tinkering with nature's own reality projection system—the human brain.

This brings us to the world of brain-computer interfaces (or BCIs). With AR contact lenses we get a nearly seamless interface with an information layer. Add in haptic gloves and the simulation starts to feel real. Take that simulation off the street and into a room, layer in both photons and ultrasound, and the experience gets even more immersive. But with BCI, we're creating reality in the exact same way that we normally create reality—with our brains.

Originally developed to help patients with "locked in syndrome" communicate, some versions of BCI use electroencephalogram (EEG) sensors to read brain waves through the scalp, enabling a hands-free, mind-controlled content interface. Already, we've seen EEG-based BCI devices start to invade gaming. Studies have demonstrated the technology in traditional arcade titles such as Tetris and Pac-man, and multiplayer games like World of Warcraft. As a result, we now have new BCI-specific games like MindBalance and Bacteria Hunt.

In 2017, researchers at the University of Washington took this farther, announcing BrainNet, the first brain-to-brain communication network that allows multiple parties to interact via their thoughts. By using electroencephalograms (EEGs) to "read" brain signals and transcranial magnetic stimulation (TMS) to "write" brain signals, participants were linked together to play a modified game of Tetris. They communicated and collaborated only through EEG, TMS, and a set of blinking lightbulbs, marking the birth of a new kind of "hive-mind gaming" and a frontier we're only beginning to explore.

We're also seeing BCI venture out of gaming and into traditional cinema. In May of 2018, English artist and director Richard Ramchurn released a twenty-seven-minute short film entitled *The Moment*. Custom-built to be watched while wearing a relatively cheap ($100) EEG headset, the contents of the film—the scenes, music, and animation—change every time you watch it, based entirely on what's going on in your head.

BCIs mean that entertainment can be not just customized for our mood, it can be customized for our brains. This is a direct computer-to-cortex connection. While the development of this connection probably exceeds the ten-year time horizon we've been focusing on, it's worth pointing out what this means for content creators. Sooner or later, media companies and neuroscience labs will begin to merge, the product of converging exponentials leading to converging markets leading to a completely unrecognizable entertainment landscape.

CHAPTER EIGHT

The Future of Education

The Quest for Quantity and Quality

The same technology converging on our movie theaters is heading for our classrooms—and just in time. From a macroscopic perspective, education has two main issues: quantity and quality. On the quantity side, we face a catastrophic shortage. In America today, we're in need of 1.6 million teachers. Globally, the picture is worse. By 2030, UNESCO estimates that the number of teachers needed will be a shocking 69 million. As a result, 263 million children worldwide currently lack access to basic education.

On the quality side, we face equally difficult challenges. Our modern educational system is anything but. It's an institution created in another time for the needs of a different world. In the mid-eighteenth century, in many ways riding the railroad across America, we spread an industrialized educational system designed to produce a standardized product. Heralded by the bell, students moved from one "learning station" to the next, while standardized tests ensured quality control—young minds well prepared for the needs of society. What were those needs? Back then, obedient factory workers.

Consider that hallmark of industrial education: the sage on the stage. This one-size-fits-all model dates to an era when great teachers and good schools were a scarce resource. While economical, a teacher preaching to a classroom filled with students tends to divide pupils into two dispirited groups: those who are lost and those who are bored.

This problem has been compounded by quality control gone wild, teachers being forced to teach the test, students standardized like never before. Sadly, what we're actually testing is a very narrow bandwidth of skills, many of which have nothing to do with the needs of adult life. Point of fact: When was the last time you factored a polynomial?

Batch processing children is both an industrial hangover and an educational disaster because of basic biology. Everyone is wired differently. Some of this is nature, some nurture, but the end result is the same: We're individuals, and there's no standard set of engaging experiences that can maximize learning for all. Put these problems together, and this helps explain why a 2015 study by the US Department of Education found some seven thousand students drop out of high school every day, or one student every twenty-six seconds. That's 1.2 million students a year, with more than half of those dropouts citing *boredom* as the number one reason they left.

But converging technology offers a host of new solutions to the challenges of quality and quantity. Every technology that's currently making an impact on entertainment is doing double duty in education, meaning, as we'll see in a moment, one-size-fits-all is no match for the app store.

One Billion Android Teachers Per Year

In 2012, Nicholas Negroponte, the founder of MIT's Media Lab, dropped off a bunch of solar charging systems and a pile of Motorola Xoom tablets in a pair of remote Ethiopian villages. The tablets were preloaded with basic learning games, movies, books, and the like, then sealed inside of boxes. Rather than being handed to the adults, the sealed boxes were given directly to children. These kids could neither read nor write. They had never seen this kind of technology before. Nobody was given instructions. Negroponte's question was simple: What happens next?

For decades, Negroponte has been trying to answer this question. He's been the leading voice for an unusual idea: Children, armed only with a laptop loaded with educational apps and games, could teach themselves to read and write while also learning how to navigate the internet.

Years earlier, to advance this cause, he'd founded the nonprofit One Laptop per Child, with the goal of building a $100 tablet computer that could be put into the hands of children in need. Still, questions remained: Was a cheap tablet sufficient to solve this problem? How much teaching and direction would kids actually need? Could children teach themselves simply by playing with the apps and games?

The Ethiopian experiment was meant to answer those questions—which it did, and then some. "I thought the kids would play with the boxes," Negroponte told the *MIT Review*. "Within four minutes, one kid not only opened the box, [but also] found the on-off switch . . . [and] powered it up. Within five days, they were using forty-seven apps per child, per day. Within two weeks, they were singing ABC songs in the village, and within five months, they had hacked [the] Android [operating system]."

Certainly, learning to read by computer is not a new idea. In *Abundance*, we explored research conducted by Sugata Mitra, a professor of education technology at Newcastle University. Mitra's work shows that functional illiteracy isn't a barrier to computer literacy. In his studies, children in India's slums were given access to a net-connected computer. Very quickly, they learned to use the equipment, surf the Web, and teach themselves the basics of reading and writing.

Negroponte's Ethiopia experiment went further. What excited the One Laptop per Child team was how the tablets unlocked self-directed learning and creativity, and more importantly, how technologically sophisticated the kids had to become, on their own, to unlock those skills in the first place. "The kids had completely customized the desktop," Ed McNierney, the nonprofit's CTO, told the *MIT Review*, "so every kid's tablet looked different. We had installed software to prevent them from doing that. And the fact they worked around it was clearly the kind of creativity, the kind of inquiry, the kind of discovery, that we think is essential to learning."

In 2017, the XPRIZE decided to take things to the next level, launching the $15 million Global Learning XPRIZE. Principally funded by Elon Musk, in partnership with Google, the prize was a software development challenge aimed at the 263 million children in the world without access to school. To claim it, a team had to develop Android-based software that would allow a child to quickly self-educate with nothing more than a tablet—that is, learn the basics of reading and writing (in Swahili, as Tanzania was where the winning software would be tested) and math in under eighteen months.

The competition attracted about seven hundred teams from around the world. Nearly two hundred delivered software, and out of this pool, five finalists were selected and each received a million dollars for their software to be loaded onto some five thousand Pixel C tablets donated by Google. In partnership with the World Food Program, XPRIZE identified some twenty-four hundred illiterate children in 167 different ultra-remote villages in Tanzania. These villages had neither schools nor literate adults. They then installed solar chargers (to charge the tablets), pretested the children (to later benchmark progress), and distributed the tablets.

In May of 2019, two teams split the final $10 million purse, Kitkit School from South Korea and Onebillion from Kenya. Both had created software that, in an hour a day, produced an education equivalent to what those children would have received attending a Tanzanian school on a full-time basis. Per the rules of the competition, the software produced by all five finalists, including the two winning teams, has been open-sourced (it's available for free on GitHub).

For this software to become a real weapon in the fight against illiteracy, there is also the issue of getting a tablet into the hands of any child in need (or adult, for that matter). But that's the real goal of the prize. If this self-education software comes pre-installed on every Android phone and tablet, then—when you're ready to replace your device—you can instead donate it to charity. You help the environment by recycling, and you help society by empowering a child. In a very real sense, you're donating a teacher. And with over a billion Android handsets manufactured each year, this software could make a serious dent in what has to be the single greatest squandering of talent in history—the 263 million young minds that need our help.

The Ultimate Field Trip

History class, 2030. This week's lesson: Ancient Egypt. The pharaohs, the queens, the tombs—the full Tut.

Sure, you'd love to see the pyramids in person. But the cost of airfare? Hotel rooms for the entire class? Taking two weeks off from school for the trip? None of these things are feasible. Yet, even if you could go, you couldn't go. Many of Egypt's tombs are closed for repairs, and definitely off-limits to a group of teenagers.

Not to worry, VR solves these problems.

In regular reality, Queen Nefertari's resting place sits in the Valley of the Queens—not that civilians ever get to look inside. To preserve the relics, the tomb has been shuttered to the public for decades. In VR world though, you and your classmates can easily visit the burial chamber, trace the hieroglyphics, even check out her sarcophagus up close. You also have a world-class Egyptologist as a tour guide: "If you turn your attention to the filigree at the back of the tomb, you will notice a sculpture of Osiris, the Egyptian god of . . ."

But turning your attention to the back of the tomb doesn't require waiting until 2030. In 2018, Philip Rosedale and his team at High Fidelity pulled off this exact virtual field trip. First, they 3-D laser scanned every square inch of Queen Nefertari's tomb. They also shot thousands of high-resolution photos of the burial chamber. By stitching together more than ten thousand photos into a single vista, then laying that vista atop their 3-D scanned map, Rosedale created a stunningly accurate virtual tomb. Next, he gave a classroom full of kids HTC Vive VR headsets. Because High Fidelity is a social VR platform, meaning multiple people can share the same virtual space at the same time, the entire class was able to explore that tomb together. In total, for their fully immersive field trip to Egypt: zero travel time, zero travel expenses.

This was a rich learning experience for the kids who took the trip. Research shows that multi-sensory learning trumps other forms—even if we do that learning in VR. This means the technology allows us to create an infinite variety of immersive high quality teaching environments. Yet this is only where we are today.

Tomorrow? Well, many experts think education could be VR's killer app. More likely, it will be a combination of VR and AI. Here's one reason why: Remember virtual Tony Robbins? The same neural nets that enable Lifekind to duplicate the renowned life coach allow us to duplicate anyone. Want to check out ancient Greece? Not only do you get every Doric column, you also get a bearded gent in a white toga greeting you with, "Hello, I'm Plato, let's tour my academy."

As cool as it might sound to learn ethics from the guy who invented ethics, VR can actually take this farther. Jeremy Bailenson, the director of Stanford's Virtual Human Interaction Lab whom we met in Chapter Three, has spent the past sixteen years studying VR's ability to expand empathy, the emotional foundation of ethics. In that time, he's discovered that VR can quickly and significantly shift our attitudes and actions toward everything from homelessness to climate change to racial prejudice. Spend time in VR world as an elderly homeless woman and the amount of empathy you feel for the homeless will significantly increase—and that increase remains once you exit the VR world. The technology doesn't just change how we feel and act in the virtual, but how we feel and act in the actual. In other words, VR unlocks the possibility of an entirely different kind of moral education.

Nor is empathy the only emotion VR appears capable of training. In research conducted at USC, psychologist Skip Rizzo has had considerable success using virtual reality to treat PTSD in soldiers. Other scientists have extended this to the full range of anxiety disorders. When you put all of this together, VR, especially when combined with AI, has the potential to facilitate a top-shelf traditional education, plus all the empathy and emotional skills that traditional education has long been lacking.

Most crucially, when AI and VR converge with wireless 5G networks, our global education problem moves from the nearly impossible challenge of recruiting teachers and funding schools for the hundreds of millions in need, to the much more manageable puzzle of building a fantastic virtual educational system that we can give away for free to anyone with a headset. It's quality and quantity on demand.

School 2030

It's 2030 and school is in session—only what does school 2030 actually look like? Turns out, our first glance at that future actually arrived in 1995, when science-fiction author Neal Stephenson published the novel *The Diamond Age*. This coming-of-age story is set in a neo-Victorian future where nanotechnology and AI are woven into the fabric of everyday life, and education is handled by the book—that is, by the *Young Woman's Illustrated Primer*.

The primer is an AI-driven, individually customized learning companion disguised as a book. The book answers questions in a contextually relevant and engaging fashion. Packed with sensors that monitor everything from energy levels to emotional state, the primer creates a rich learning environment aimed at producing a specific transformation. Rather than molding children to the needs of society, the primer has more humanist aims: to produce strong, independent, empathetic, and creative thinkers.

As it turns out, Neal Stephenson is now the chief futurist at Magic Leap, helping use augmented reality to birth his illustrated primer, version 1.0. Magic Leap's technology allows you to place holograms in the world around you. Concepts that are difficult to visualize via a 2D screen—such as human anatomy—come alive in this 3-D world. Imagine a virtual autopsy, being able to strip away layers of skin or muscle inside a navigable operating room. The experience of learning in a 3-D environment is a rich one, making it much more likely to cross the bridge from short-term into long-term memory.

But the real magic of augmented reality is that it extends the classroom into the world. With the mix of AR and AI, every walk becomes a history lesson. Amble the streets of Manhattan, for example, and you can see the buildings as they were a century ago, complete with holographic Victorians serving as virtual historians.

Of course, AR alone doesn't get us to the *Primer*, but if we couple it to ongoing convergences, the picture becomes clearer. Today's AI revolution gives us another component, the ability to create individually customized learning environments. Add in sensors that respond to neurophysiological data—so students can, for one example, keep themselves in a growth mindset (which research shows is needed for learning), or, to offer a second example, push themselves toward a flow state (which research shows can amplify learning). Put it all together and we start to see a very different future, one of distributed, individually customizable, accelerated learning environments. So what is school like in 2030? Well, what would you like to learn today?

CHAPTER NINE

The Future of Healthcare

Martine and the Moonshots

It was the news no one wants to receive.

In 1992, Martine Rothblatt's daughter was told she had fewer than five years to live. The doctors called it pulmonary hypertension, a rare lung disorder that historically infects about two thousand Americans at any one time. Yet it's easy to misinterpret the smallness of this number—the disease itself is a ruthless killer, claiming just about every life it touches. The small number of people living with pulmonary hypertension is a testament to its virulence, not its frequency. Either way, her daughter was dying, so Martine decided to square off against this killer.

The doctors told her that her quest was impossible. There were a lot of doctors. Also, a lot of time between doctor's visits—mostly spent in a medical library. Martine had a system: Find an article in a journal about pulmonary hypertension, backtrack the terminology to a college textbook, backtrack the key ideas to a more general high school textbook, and repeat.

And repeat.

Martine doesn't remember the moment she decided to tackle the ultimate moonshot—cure an incurable disease in less time than it would take her daughter to die from the disease—but that's par for the course. By the time she got interested in pulmonary hypertension, Martine Rothblatt was already two moonshots in to what is currently a seven-moonshot career. And counting.

Today, Martine Rothblatt is one of the highest paid female CEOs in America. How she got there though? That's the more interesting story.

Martine began her life as Martin, a Jewish boy from a Hispanic neighborhood in Chicago. He grew up to be, at least for a while, nothing spectacular. First a college dropout, next a globetrotting backpacker. But a chance encounter with a NASA tracking system in the Seychelle Islands gave him a crazy idea: unite the world through satellite communications.

Martin, like Martine, was a go-getter. His Seychelles vision led to graduate school at UCLA for dual degrees in law and business. He parlayed those into an expertise in space law, which laid the foundation for a series of space-based communications companies. These include both the world's first global satellite radio network, and Sirius XM, still the satellite radio category killer, which Rothblatt cofounded in 1990.

In the middle of this, Martin got married, had a daughter named Jenesis, got divorced, got remarried, had two more children, then decided he was a she trapped in the wrong body. So Martin embarked on her second moonshot, sex reassignment surgery, and became Martine, staying married to the same woman. Actually, they stayed happily married.

But that's when Jenesis got sick.

Martine cashed out of Sirius. She plunged that money into hunting for a cure. Eventually, this led her to an orphan drug for pulmonary hypertension. Glaxo owned the patent, but they'd shelved it. Martine built a team of scientists and managed to license the "drug"—though we're using that word euphemistically. What she actually got from Glaxo was a small Baggie filled with a few tablespoons of white powder that—in tests with rats—had showed a little promise a long time ago.

Still, United Therapeutics was born.

A hundred top chemists said the patent would never become a

medicine, but three years later, when Martine's daughter was literally taking her last breaths, that medicine hit the market. Today, Jenesis is in her mid-thirties, the drug that saved her life generates a billion and a half dollars a year in revenue for United Therapeutics, and the number of patients now living with pulmonary hypertension has climbed from two thousand to forty thousand.

And if this were the end of the story, it would be a hell of a tale. Yet Martine's drug was a half measure. It managed the condition, but wasn't a cure. In fact, right now, the only cure for pulmonary hypertension—or, for that matter, pulmonary fibrosis, cystic fibrosis, emphysema, or COPD—is a lung transplant. But in the US, only two thousand lungs a year become available, while over a half-million people die of lung failure from tobacco-related ailments alone. These dire facts led to another Martine moonshot: create an unlimited supply of transplantable organs.

"We do this with cars and buildings all the time," explains Martine, "we swap out old parts for new parts and can keep things running, essentially, forever. I wanted to find a way to do this for the human body."

She took a three-pronged approach to this problem. First, to solve the problem of lung replacement, she decided not to reinvent the wheel. Today, because the lungs of the dying fill up with toxic chemicals, over 80 percent of those donated for transplant end up in trash cans. So Martine helped perfect a way to keep lungs alive outside the body, what's technically called "ex vivo lung perfusion." Already, this procedure has saved thousands of lives, but once again, she wasn't done.

Next, Martine attacked the larger problem of organ shortages through xenotransplantation. It's an old and controversial idea—harvest fresh animal organs to replace failing human ones—but issues of disease, rejection, and animal cruelty have kept it sidelined. Martine decided to push through.

Pig organs are similar to human organs, so she started there. By teaming up with Synthetic Genomics and Craig Venter, the same scienst who decoded the human genome, she made the most complete genetic map of a pig to date. Next, CRISPR knocked out all the genes that led to viruses, eliminating the dangers of disease and producing a "clean" pig. Now their latest goal is the biggest: knock out the genes that lead to organ rejection in humans. If successful, it'll mean a near infinite organ supply—albeit one that comes with a whole lot of suffering for pigs.

To combat that final problem, Martine is using cutting-edge tissue engineering techniques in an attempt to bypass animals entirely. Out of collagen, she's begun 3-D printing an artificial lung scaffold. To turn that scaffold into a living lung, she's experimenting with stem cells.

And, lastly, because it often takes too long to get an organ from its current location to a waiting patient, Martine backed Beta Technologies' flying car, with plans to use these eco-friendly vehicles to whisk newly minted organs to patients in need. Finally, at age sixty, just for the fun of it, she became a helicopter pilot herself, and then, in a vehicle designed by her company, set a world record for speed in an electric helicopter. All of which is to say, by sometime in 2028 or so, Martine believes that death by organ failure will become a problem to be managed rather than a sad fact of life. And we have seven moonshots' worth of reasons to believe her.

Turning Sick Care into Healthcare

It probably goes without saying that Martine Rothblatt's outsider assault on the healthcare industry was aided and abetted by converging exponentials. CRISPR, genomics, stem cells, 3-D printing, electric vehicles—the list goes on. But it's helpful to remember that even though Martine's story is a testament to what determination and technology now make possible, it's only one story. There are thousands more, maybe not quite as extraordinary, but just as impactful.

When it comes to healthcare, the system itself is often sicker than the patients. Even the terminology is misleading. Today, going to the doctor is about *sick* care more than *health*care. It's reactive, not proactive. Doctors make after-the-fact interventions, fighting a rearguard battle that's often inefficient, overpriced, and in certain cases, downright surreal. In the US, for example, fear of liability has doctors spending $210 billion per year on procedures patients don't need.

The research side isn't any better. Out of every five thousand new drugs introduced, only five make it to human testing, and only one of those is actually approved. This is why the average medicine takes twelve years to get from lab to patient, at a cost of $2.5 billion, and Americans spend an average of $10,739 per person per year on healthcare—more than any other country on Earth. If nothing changes, by 2027, this single industry will consume nearly 20 percent of the US GDP.

But plenty is changing. The scope of this story is huge. If we wanted, we could fill multiple volumes with all that's unfolding. To keep things manageable, we're focusing on six shifts, four that are technological and two paradigm shifts.

On the technological front, every step in the medical treatment train is being reinvented. On the front end, the convergence of sensors, networks, and AI is upending medical diagnostics. In the middle, robotics and 3-D printing are changing the nature of medical procedures. On the back end, AI, genomics, and quantum computing are transforming medicines themselves.

Concurrently, as a result of these convergences, two major paradigm shifts are under way. The first is the shift from sick care to healthcare, from a system that is retrospective, reactive, and generic, to one that is prospective, proactive, and personalized.

The next is a change in management. For most of the last century, the healthcare industry was an uneasy partnership between big pharma, big government, and the full spectrum of doctors, nurses, and trained medical professionals. Now we're witnessing an invasion. Many of the big technology companies are getting into this game, all intent on making an impact. "If you zoom out into the future," Apple CEO Tim Cook recently said (in that same interview with the *Independent* where he talked about the potential of AR), "and ask what was Apple's greatest contribution to mankind, it will be about health."

Racing Apple are Google, Amazon, Facebook, Samsung, Baidu, Tencent, and others. As we shall see in a moment, all of these companies have three clear advantages over the establishment: They're already in your home, into artificial intelligence, and experts in collecting and analyzing your data. While it remains an open question whether we want to turn our healthcare over to the big technology companies, what is certain is that these three advantages are fundamental to detecting diseases early enough to make a difference, which is definitely the first step in turning sick care into healthcare.

DIY Diagnostics

On a wintery Wednesday in January 2026, you're being watched. Carefully watched. Technically, you're asleep in your bed, but Google's home assistant knows your schedule. Thanks to your Oura ring, it also knows you've just completed a REM cycle and are now entering Stage 1 sleep—making it the perfect time to wake you up.

A gentle increase in the room's lighting simulates the sunrise, while optimized light wavelengths maximize wakefulness and improve mood. By the time you've gone through your bathroom rituals—toilet, toothbrush, etc.—you realize mood isn't the problem. It's that tightness in your joints, the chill in your bones.

Are you getting sick?

The NIH released their universal flu vaccine a few months back, but you didn't have time for the shot. Now you're wondering if skipping it was a mistake.

But no need to wonder.

"Hey Google, how's my health this morning?"

"One moment," says your digital assistant.

It takes thirty seconds for the full diagnostic to run, which is pretty good considering the system deploys dozens of sensors capturing gigabytes of data. Smart sensors in toothbrush and toilet, wearables in bedding and clothing, implantables inside your body—a mobile health suite with a 360-degree view of your system.

"Your microbiome looks perfect," Google tells you. "Also, blood glucose levels are good, vitamin levels fine, but an increased core temperature and IgE levels . . ."

"Google—in plain English?"

"You've got a virus."

"A what?"

"I ran through your last forty-eight hours of meetings. It seems like you picked it up Monday, at Jonah's birthday party. I'd like to run additional diagnostics, would you mind using the . . ."

Well, take your pick. Alphabet's healthcare division, called Verily Life Sciences, is developing a full range of internal and external sensors that monitor everything from blood sugar to blood chemistry. And that's just Alphabet. The list of once multimillion-dollar medical machines now being dematerialized, demonetized, democratized, and delocalized—that is, made into portable and even wearable sensors—could fill a textbook.

Consider the spectrum of possibilities. On the whiz-bang side, there's Exo Imaging's AI-enabled, cheap, handheld ultrasound 3-D imager—meaning you will soon be able to track anything from wound healing to fetus growth from the comfort of your home. Or former Google X project leader Mary Lou Jepsen's startup, Openwater, which is using red laser holography to create a portable MRI equivalent, turning what is today a multimillion-dollar machine into a wearable consumer electronics device and giving three-quarters of the world access to medical imaging they currently lack. Yet simpler developments might be more revolutionary.

In less than two decades, wearables have gone from step-counting first-generation self-trackers to Apple's fourth-generation iWatch that includes an FDA-approved ECG scanner capable of real-time cardiac monitoring. Or Final Frontier Medical Devices' DxtER, the winner of the $10 million Qualcomm Tricorder XPRIZE, a collection of easy-to-use, noninvasive medical sensors and a diagnostic AI accessible via app. DxtER reliably detects over fifty common ailments.

All of these developments point toward a future of always-on health monitoring and cheap, easy diagnostics. The technical term for this shift is "mobile health," a field predicted to become a $102 billion market by 2022. Step aside, WebMD. The idea here is to put a virtual doctor, on demand, in your back pocket.

And we're getting close. Riding the convergence of networks, sensors, and computing, AI-backed medical chatbots are now flooding the market. These apps can diagnose everything from a rash to retinopathy. And it's not just physical ailments. Woebot is now taking on mental health, delivering cognitive behavioral therapy via Facebook Messenger to patients suffering from depression.

So where are these trends actually headed?

Take Human Longevity Inc. (HLI), one of the companies Peter cofounded. HLI offers a service called "Health Nucleus," an annual, three-hour health scan consisting of whole genome sequencing, whole body MRI, heart and lung CT, electrocardiogram, echocardiogram, and a slew of clinical blood tests—essentially the most complete picture of health currently available.

This picture is important for two reasons. The first is early disease detection. In 2018, Human Longevity published stats on their first 1,190 clients. Nine percent of their patients uncovered previously undetected coronary artery disease (the number one killer in the world), 2.5 percent found aneurysms (the number two killer in the world), 2 percent saw tumors—and so forth. In total, 14.4 had significant issues requiring immediate intervention, while 40 percent found a condition that needed long-term monitoring.

The second reason this is important? Everything Human Longevity is measuring and tracking via half-day annual visits will soon come to you on demand. Thanks to always-on, always watching sensors, your smartphone is about to become your doctor.

Reading, Writing, and Editing the Code of Life

For a decade, experts have been trumpeting personalized genomics as a healthcare revolution. When we understand your genome, the thinking goes, we'll know how to optimize "you." We'll know the perfect foods, the perfect drugs, and the perfect exercise regimen, just for you. We'll know the types of gut flora best suited for your microbiome, the supplements that best commingle with your physiology. You'll learn the diseases to which you're most susceptible, and, more importantly, how to prevent them. Or so the story goes. . . .

In 2017, Jason Vassy, a professor of medicine at Boston's Brigham and Women's Hospital, decided to take a closer look at that story. A hundred patients were recruited. Half had their DNA screened; the other half answered questions about family medical history, which is the standard method for establishing genetic risk. In the results, Vassy wanted to compare overwhelm and anxiety versus real world usefulness. Critics of personalized genomics worry about information overload for doctors, needless anxiety for patients, and expensive and unnecessary follow-up testing for both. But that's not what Vassy discovered.

Instead, according to results published in the *Annals of Internal Medicine*, there was no trace of these concerns. What did happen was that 20 percent of the patients who had their DNA screened discovered rare, life-threatening conditions that required immediate action. Once again, in results very similar to what Health Nucleus uncovered, lives were saved.

Yet the more important result comes not from any one patient being screened, but from their combined aggregate of genomes. The larger and more complete our genetic data sets, the more robust the preventative power of genomics will become. This is also why, in 2018, the National Institutes of Health launched their All of Us project, distributing nearly $27 million in grants to sequence a million genomes, and Harvard geneticist George Church recently founded Nebula Genomics to do something similar.

Church is also involved in the Genome Project-Write, which is the farther future—an attempt to write a human genome from scratch. If successful, GP-Write, as it is known, will help us grow transplantable organs, give us new weapons in the war on viruses and cancer, and provide cheap medicines and vaccines.

Another frontier is the use of CRISPR to edit genomes. It's still early, but progress is impressive. Researchers recently genetically engineered cocaine resistance into mice, switched off the gene responsible for Duchenne muscular dystrophy in dogs, and have begun developing personalized cancer therapies in humans. There's even work in insects. Researchers at Imperial College in London have CRISPR-created a new breed of mosquito that cannot reproduce. It's also designed to outcompete its malaria-laden siblings, making this a healthcare revolution via species-wide gene editing—and it's already happening. In late 2018, field trials were getting under way in Burkina Faso, a country plagued by malaria.

But the biggest news isn't about this company or that technique, it's that half of the thirty-two thousand most common genetic disorders are caused by an error in a single base pair—meaning one letter in the code is out of place. This may be something we can soon correct. We're not there yet, but before too long, between traditional gene therapy and CRISPR, we will soon gain the ability to purge sixteen thousand diseases from our lives. So you've got to ask yourself, if the curing of one disease is a miracle of the biblical variety, what do you call the curing of sixteen thousand?

The Future of Surgery

There's no healthcare on Mars. The Red Planet is devoid of hospitals, HMOs, and all the rest. And while none of this matters today, in the 2030s, when NASA is planning to launch the first human exploration mission to the planet, it'll be a much bigger issue. On Mars, astronauts won't just be out of network, they'll be off-world, with the nearest emergency room nine months and a gravity-assist away.

Traumatic injury keeps astronauts up at night. Not only has it never happened before—so we have zero experience with this particular space catastrophe—it's all but guaranteed to happen on Mars. Studies put the odds of a serious medical problem in space at .06 percent per person per year. On a multiyear interplanetary mission, remaining emergency free would require, well, as Elon Musk once explained: "If safety is your top goal, I would not go to Mars."

Dr. Peter Kim wants to solve this problem. An associate surgeon in chief at Children's National Medical Center in Washington, DC, Kim is part of the research team behind STAR—Soft Tissue Autonomous Robot—a robot that can already outperform surgeons on the specific task of suturing soft tissue.

Soft tissue repair is messy. It's bloody. It demands exacting precision. Since doctors vary in both levels of training and dexterity, over 30 percent of soft tissue surgeries end in complications. In space, these kinds of complications can easily become fatal, so figuring out how to perform soft tissue surgery before we start colonizing other planets is mission critical.

STAR is one of our best hopes. For starters, dexterity is standard—it comes built in. And with an AI involved, training isn't an issue. Right now, STAR sews tissue five to ten times faster than a human, and with far greater precision. Tomorrow's version will have more sophisticated force feedback and an array of multispectral cameras that can see through soft tissue. Kim wants tomorrow's system aboard the first mission to Mars to ensure that surgery in space doesn't end up looking like outtakes from *Alien*.

Yet, while critical off-world, STAR's real promise is here on Earth. There are roughly 50 million surgeries undertaken in the US each year, yet fewer than 5 percent are done robotically. However, the single most important question you can ask as you interview your surgeon is "How many times have you performed this procedure?" More importantly, "How many times have you performed it today?" Surgeons with the most practice in the widest variety of conditions produce the best results. This is why, ten years from now, when you're wheeled into the operating room and see a human doctor, your immediate response will be: "No way. I want the robot."

Dozens of surgical robots are now heading to market. There are bone-crushing robots for orthopedics already in use, with five different spinal surgery robots about to come online, and specialty bots of nearly every variety in development. Most of these are cobots—meaning the robot assists the surgeon rather than replacing the surgeon. Yet it's autonomous robots like STAR that hold the most promise. With the ability to perfectly execute routine procedures at a fraction of today's cost, robo-surgeons bring demonetization into the operating room.

Unwilling to let entrepreneurs have all the fun, big tech companies are also rushing into that room. Exhibit A: Verb Surgical, a partnership between Alphabet and Johnson & Johnson. With their fleet of cheap and much-improved surgical robots hitting the market in 2020, Verb's stated, unassuming goal is the "democratization of surgery." What does that mean in plain English? Your medical bills will get a lot smaller.

And while full-grown surgical robots have gotten much of the attention, their smaller cousins might be the more impactful development. Take the Israeli startup, Bionaut Labs. In medicine today, a great many of the issues we face are local in nature. Cancer, for example. We have lung cancer or ovarian cancer. Unfortunately, we often treat local cancers with system-wide solutions, such as chemotherapy. These system-wide approaches tend to be inaccurate, inefficient, and prone to side effects, which are three reasons why drug design is stratospherically expensive and 90 percent of all potential treatment candidates fail to make it out of the lab.

But Bionaut has built an almost microscopic robot that moves through tissue at speed—some sixty centimeters per hour—in a minimally invasive, absolutely precise manner. Guided by weak magnetic fields, these remote-controlled microbots carry diverse payloads that can be released on demand, exactly when and where needed. They're a few years away from prime time, but the plan is to use them for diagnostics, targeted medicine delivery, and minimally invasive surgery.

While both macrobots in operating rooms and microbots in our bodies will transform surgery, in our converging universe nothing ever operates on its own. AI is already in this surgical mix. It parses the blizzard of signals pouring into the ICU, aids autonomous robots as they navigate the human body, and—via cobots like the Da Vinci robot—steadies the tremors in a surgeon's hand. But, once again, it's not just AI.

3-D printing has also entered the operating room. Actually, it's been there for a while. In *Abundance*, we described how the tech was then hitting prosthetics, beginning to impact organ printing, and about to enter bionics. Today, an internet search reveals people with little training creating exceptionally functional prosthetic limbs on equipment bought at Staples. People with training, meanwhile, are crafting organs, ears, heart shunts, spinal cords, cranial plates, hip joints, and individually customized surgical tools. And with the ability to 3-D print electronics, we're making bionic body parts as well. In fact, in 2018, a team from the University of Minnesota successfully printed a rounded semiconducting material capable of turning light into patterns—which has been the main impediment to creating that ultimate spare part: a printable, bionic eye.

Cellular Medicine

Cellular medicine was a notion that first emerged in the 1990s, after the discovery of stem cells. Novel yet simple, the idea was to use these cells as weapons against illness. In the years since, this idea has come to include more cell types than just stem cells, but the treatment remains the same. Inject patients with living cells which, to a varying degree, can influence and/or revitalize an assortment of functions: regrow hair, rejuvenate tissue, kill cancer, repair cardiac damage, quiet autoimmune diseases, even increase muscle mass.

Earlier in this book we mentioned neurosurgeon and entrepreneur Bob Hariri, who helped pioneer the field of cellular medicine with his year 2000 discovery that the human placenta houses an abundant supply of stem cells—providing a noncontroversial supply of this potential treatment option.

After Hariri's company was acquired by pharma giant Celgene, he led a team of over a hundred scientists and engineers in an effort to turn placental stem cells into real medicines. Along the way, they made two other critical discoveries. First, as people age, their supply of stem cells rapidly diminishes, a process known as "stem cell exhaustion" (that we'll explore in depth in our next chapter). Second, the placenta doesn't just contain stem cells, but also houses immunological cells such as natural killer cells and T-cells, both of which are critical in the body's natural ability to fight cancer—as long as they recognize the danger.

Normally our immune system destroys cancer cells at very early stages of development. But, as we age, cancers can pile up. Some go undetected, and that's when the situation gets dangerous. To attack that danger, we've invented a new kind of therapy known as CAR-T (for chimeric antigen receptor T-cell) therapy. In this approach, a patient's white blood cells are collected and their T cells are separated out and genetically reengineered to target and kill specific cancer cells. These reprogrammed cells are then injected back into the patient, effectively becoming a kind of heat-seeking cancer missile.

Unfortunately, this isn't cheap.

In 2017, when the first CAR-T therapies were released, the price was roughly half-a-million dollars per patient. Since every patient's CAR-T cells have to be individually weaponized, the problem was how to manufacture this medicine at scale. In 2018, Celgene spun out their cellular medicine division with Hariri at the helm, and by using placental-derived immunological cells, this new company, Celularity, has created a one-size-fits-all version of the medicine. Rather than a bespoke therapy, Celularity can create CAR-T in bulk and at speed, getting drugs to patients in hours after diagnosis, rather than the weeks it now takes.

Celularity scientists have also figured out how to weaponize placental natural killer cells (pNK cells), genetically modifying them into CAR-NK cells to enhance their tumor-targeting abilities. And like placental CAR-T cells, placental CAR-NK cells can be made into a one-size-fits-all medicine, giving us the ability to make cancer treatments available to the masses. This is the most important point. Cancer is the number two killer in the world, and placentas are abundant. There are over a hundred million births per year, and 99 percent of those placentas are thrown away. Saving this supply gives us the potential to manufacture these medicines cheaply and at scale.

The Future of Drugs

Traditionally, if a pharmaceutical company wants to create a new medicine, they have two choices. Either they comb enormous medical libraries for potential candidates or send an expedition to an exotic locale to hunt for naturally occurring possibilities—like some rare tree bark that has anti-cancer properties. Neither is a sure thing, both take years of effort, and that's barely the starting line. Once candidates are identified, they're next analyzed, synthesized, and more years pass. Finally, these discoveries are tested, initially in animals, then in small groups of humans, eventually in large groups of humans. In short, drug discovery is a long, slow war.

It really is a war. The body count is high. Ninety percent of all drug possibilities fail. The few that do succeed take an average of ten years to reach the market and cost between $2.5 billion and $12 billion to get there. But computer scientist turned biophysicist Alex Zhavoronkov believes he's found a shortcut.

Around 2012, Zhavoronkov started to notice that artificial intelligence was getting increasingly good at image, voice, and text recognition. He knew that all three tasks shared a critical commonality. In each, massive datasets were available, making it easy to train up an AI. Similar datasets were present in pharmacology. So, back in 2014, Zhavoronkov started wondering if he could use these datasets and AI to significantly speed up the drug discovery process.

He'd heard about a new technique in artificial intelligence known as generative adversarial networks (or GANs). By pitting two neural nets against one another (adversarial), the system can start with minimal instructions and produce novel outcomes (generative). Researchers had been using GANs to do things like design new objects or create one-of-a-kind, fake human faces, but Zhavoronkov wanted to apply them to pharmacology. He figured GANs would allow researchers to verbally describe drug attributes: "The compound should inhibit protein X at concentration Y with minimal side effects in humans," and then the AI could construct the molecule from scratch.

To turn his idea into reality, Zhavoronkov set up Insilico Medicine on the campus of Johns Hopkins University in Baltimore, Maryland, and rolled up his sleeves. "It took us three years of hard work to develop a system that researchers could actually interact with in this way," he explains. "But we pulled it off, and this has allowed us to reinvent the drug discovery process."

Instead of beginning their process in some exotic locale, Insilico's "drug discovery engine" sifts millions of data samples to determine the signature biological characteristics of specific diseases. The engine then identifies the most promising treatment targets and—using GANs—generates molecules (that is, baby drugs) perfectly suited for them. "The results are an explosion in potential drug targets and a much more efficient testing process," says Zhavoronkov. "AI allows us to do with fifty people what a typical drug company does with five thousand."

The results have turned what was once a decade-long war into a month-long skirmish. In late 2018, for example, Insilico was generating novel molecules in fewer than forty-six days, and this included not just the initial discovery, but also the synthesis of the drug and its experimental validation in computer simulations.

Right now, they're using the system to hunt for new drugs for cancer, aging, fibrosis, Parkinson's, Alzheimer's, ALS, diabetes, and many others. The first drug to result from this work, a treatment for hair loss, is slated to start Phase I trials by the end of 2020. They're also in the early stages of using AI to predict the outcomes of clinical trials in advance of the trial. If successful, this technique will enable researchers to strip a bundle of time and money out of the traditional testing process.

Beyond inventing new drugs, AI is also being used by other scientists to identify new drug targets—that is, the place a drug binds to in the body and another key part of the drug discovery process. Between 1980 and 2006, despite an annual investment of $30 billion, researchers only managed to find about five new drug targets a year.

The trouble is complexity. Most potential drug targets are proteins, and a protein's structure—meaning the way a 2D sequence of amino acids folds into a 3-D protein—determines its function. But a protein with merely a hundred amino acids (a rather small protein) can produce a googol-cubed worth of potential shapes—that's a one followed by three hundred zeroes. This is also why protein-folding has long been considered a really hard problem for a really big supercomputer.

Back in 1994, to monitor this supercomputer protein-folding progress, a biannual competition was created. Until 2018, success was fairly rare. But then the creators of DeepMind turned their neural networks loose on the problem. They created an AI that mines enormous datasets to determine the most likely distance between a protein's base pairs and the angles of their chemical bonds—aka, the basics of protein folding. They called it AlphaFold.

On its first foray into the competition, contestant AIs were given forty-three protein-folding problems to solve. AlphaFold got twenty-five right. The second-place team managed a meager three.

If we couple AlphaFold's progress to Insilico's GANs and add in the anticipated breakthroughs in quantum computing—another technology being aimed at drug discovery—we're not far from a world where individually customized medicine will move from science fiction to the standard of care. And don't blink, because as radical a shift as this may seem, none of it includes the breakthroughs happening in the adjacent field of longevity.

CHAPTER TEN

The Future of Longevity

The Nine Horsemen of Our Apocalypse

Earlier in this book, we explored how extending the healthy human lifespan will have a significant impact on the rate of change in the world. The equation was fairly simple—longer lives means more time spent at our productive best which means more innovation. But what we didn't cover in any detail was how this would happen. Here, on the back end of our exploration of the related field of healthcare, we want to turn our attention to this question of longevity, seeing how the forces of convergence are rewriting the rules in the race between technology and mortality.

And the place to start: mortality itself, the life clock known as aging. "Aging is not just a running down of the system," explains longevity researcher and director of the National Institutes of Health, Francis Collins. "It is a programmed process. Evolution probably had an investment in having the lifespan of a particular species not go on forever. You've got to get the old folks out of the way so the young ones have a chance at the resources."

To get the old folks out of the way, evolution devised a fail-safe: planned obsolescence, otherwise known as aging. It's a redundant plan.

Scientists now believe there are nine main "causes" of our decline, the nine horsemen of an internal apocalypse. We're going to spend the rest of this chapter investigating the strategies being deployed to defeat this decline. Yet, before we do that, we first need to meet these horsemen, and explore the fundamental question they answer: What, exactly, is killing us?

1. ***Genomic Instability:*** DNA doesn't always replicate according to plan. Typically, these errors in gene expression get caught and corrected, but not always. Over time, these misfires build up, causing our body to wear down—meaning genetic instability leads to genetic damage leads to a limit on lifespan. Think of it as a broken copy machine, except, instead of producing unreadable pages, our broken genetic copier produces diseases like cancer, muscular dystrophy, and ALS.

2. ***Telomere Attrition:*** At the heart of a cell, DNA is packed into threadlike structures called chromosomes. Chromosomes are capped by telomeres, or short snippets of DNA repeated thousands of times. These repetitions act as barriers—like bumpers on a car—designed to protect the core of the chromosome. But as DNA replicates, telomeres get shorter. At a critical shortness threshold, the cell stops dividing, and we become much more susceptible to disease.

3. ***Epigenetic Alterations:*** Nature impacts nurture. Over the course of a lifetime, factors in our environment can change how our genes express, sometimes for the worse. Exposure to carcinogens in the environment can silence the gene that suppresses tumors, for example. These cells start to grow uncontrollably and cancer is the result.

4. ***Loss of Proteostasis:*** Inside a cell, proteins run the show. They transport materials, send signals, switch processes on and off,

and provide structural support. But proteins become less effective over time, so the body recycles them. Unfortunately, as we age, we can lose this ability. The trash collector goes on strike and we suffer a toxic buildup of proteins that can, for example, lead to diseases such as Alzheimer's.

5. ***Nutrient Sensing Goes Awry:*** The human body relies on over forty different nutrients to stay healthy. For everything to work perfectly, cells need to be able to recognize and process each of these. But this ability breaks down as we get older. For example, one reason people gain weight as they age is that our cells can no longer properly digest fat. And one reason we die is that this impacts the insulin and IGF-1 pathway and can result in diabetes.

6. ***Mitochondrial Dysfunction:*** Mitochondria are power plants. By converting oxygen and food into energy, they provide the basic fuel for our cells. But performance declines over time. The result is free radicals, a damaging form of oxygen that mangles DNA and proteins and leads to many of the chronic illnesses associated with aging.

7. ***Cellular Senescence:*** As cells undergo stress, they occasionally become "senescent," both losing their ability to divide and, simultaneously, becoming resistant to death. These "zombie cells" can't be removed from the body. They build up over time, infect neighboring cells, and ultimately create a zombie apocalypse of inflammatory debilitation.

8. ***Stem Cell Exhaustion:*** As we age, our supply of stem cells plummets, in certain cases by a ten thousandfold decline. Worse, the ones we do manage to hang on to become far less active. This means that the body's internal tissue and organ repair system loses its ability to do its job.

9. ***Altered Intercellular Communication:*** For the body to function properly, cells need to communicate. This happens constantly, with messages flowing through our bloodstream, immune system, and endocrine system. Over time, signals get crossed. Some cells become unresponsive, others become inflammation-producing zombie cells. This inflammation blocks further communication. Once this happens, messages can't get through and the immune system can't find pathogens.

And now that we know what's killing us, let's see what just might be saving us.

Longevity Escape Velocity

Want to win a Nobel Prize? Study worms. And don't just study any worm. Study the roundworm, Caenorhabditis elegans, or, as her friends call her, *C. elegans*.

And this worm has a lot of friends.

Six scientists have already taken home Swedish gold for their work on the creature. As a result, *C. elegans* was the first organism to have its genes sequenced, its whole genome screened, and its connectome, the wiring diagram of the brain's neurons, mapped. But despite this celebrated history, many feel the roundworm's greatest contribution is still to come, as *C. elegans* is also the first animal to go head-to-head with death—and win.

In a petri dish, *C. elegans* lives about twenty days. Back in 2014, a group of NIH scientists at the Buck Institute for Research on Aging decided to try and increase that number. Previous research had shown there were two ways to have an impact. Knocking out a gene named rsks-1 increases lifespan by six days; knocking out daf-2, meanwhile, extended it by twenty days. But what happened, these researchers wanted to know, if they knocked out both genes at once?

"Taking an educated guess, [the researchers] estimated that such double-mutants might live about forty-five days," wrote NIH director Francis Collins, who funded the work. "But, to their surprise, when they actually created such worms, some of the critters were still alive and squirming at a hundred days. That's an amazing five-fold increase in worm lifespan—the equivalent of four-hundred-year-old humans."

Applying this same process to human lifespan—that's exactly what's at the heart of the field of longevity. Genetics, of course, play a critical role. Building out on this earlier work on *C. elegans*, other researchers have since identified over fifty more genes that seem to trigger age-related decline. Five of these genes seem especially key, as removing any of them produces a 20 percent boost in lifespan.

But it's not just genetics. Martine Rothblatt's mission to produce an endless supply of human replacement parts is also crucial to longevity. As is the democratization of surgery being provided by robotics, and the drug discovery work being done by AIs and quantum computers. But the point isn't this or that technique, it's the combinatory power of all of these approaches that are leading us in a very new direction.

The old direction was our thirty-year lifespan, which held constant from the Paleolithic Age to the front end of the Industrial Revolution. During the twentieth century, marvels such as antibiotics, sanitation, and clean water extended our average age to forty-eight years by 1950, then to seventy-two years by 2014. But these days, Ray Kurzweil and longevity expert Aubrey de Grey have begun talking about "longevity escape velocity," or the idea that soon, science will be able to extend our lives by a year for every year we live. In other words, once across this threshold, we'll literally be staying one step ahead of death.

Kurzweil thinks this threshold is about twelve years away, while de Grey puts it thirty years out. Why should we believe them? One basic fact: You can't take it with you. All the money in the world is useless in the grave. So how much would the wealthy pay for an extra healthy decade or two or three? A lot. This helps explain the increasing investments being made into anti-aging technologies, with Google's Calico—an acronym standing for "California Life Company"—being perhaps the most visible example. More critical, while a longer life for the wealthy might not seem that worthwhile of a goal, as we've seen with every other accelerating technology, it isn't long before the benefits become demonetized and democratized. And this means that possibly you, and definitely your children, will have the potential to tack decades onto your lives, simply because, as time passes, all of us will intercept a gaggle of anti-aging technologies along the way.

Let's turn our attention to a few of the most promising ones.

The Anti-Aging Pharmacy

Easter Island is remote. It's exotic. It's home to strange rumors and stone heads and sometimes strange rumors about stone heads. Some say the elders, with the right spells, can wake the heads from their slumber, controlling them like a giant stone army. Others say the heads themselves have the control—over your life force, with both the ability to steal it, bringing on an early death, or amplify it, conferring virility and strength on a chosen few. Then, in the mid-1960s, a small team of researchers discovered that this last bit, that conferring of strength and virility part, might be more than a rumor.

It started when the very small and very isolated community living on Easter Island decided they'd had enough. Enough smallness. Enough isolation. It was time, they decided, to build an airport.

Scientists freaked out. One of the most ecologically untainted regions of the world was about to lose its purity. In an emergency effort, an international team was rushed in to collect flora, fauna, and microbial samples, including—most critically for this story—dirt excavated from beneath one of the island's mysterious heads.

The dirt ended up in the hands of a Canadian microbiologist named Suren Sehgal, who discovered it did, in fact, contain magical powers—of the antifungal variety. Sehgal purified the compound, naming it rapamycin after the island's original name, Rapa Nui. Despite its potential, Sehgal's research money ran out, and the compound was shelved until the late 1970s, when he got enough funding to take a second look. This is when he discovered there was more magic in that dirt. Rapamycin wasn't just an antifungal, it also suppressed the immune system, giving it a lot of potential in organ transplant operations.

This potential became an industry. Rapamycin has since been used for everything from coating heart stents to ensuring that patients don't reject their new kidney. And then researchers made an even more incredible discovery about this magic dirt: Rapamycin inhibits cancer growth.

The compound blocks a protein that facilitates cell division. Do this in worms, flies, and yeast and the result isn't just cancer protection, it's longevity. This raised the next question: Would the magic work in mammals?

In 2009, NIH scientists answered that one, showing that rapamycin extends the lives of mice by as much as 16 percent. In 2014, the combination of all of these findings led Novartis to decide to test it in humans, marking the first official trial of an anti-aging compound by a major drug company. But once scientists figured out that there really was magic in that dirt, the search for other anti-aging compounds was under way.

One place this search led was our medicine cabinet, where we found a drug called metformin. The world's most common diabetes drug, metformin blocks sugar production and helps regulate insulin. But it also slows the "burn rate" of cells, defending against oxidative stress, fighting cancer, and—as we recently learned—significantly extending the lifespan of worms, mice, and rats. Does it work in humans? That remains an open question, but researchers are trying to find out.

While rapamycin and metformin protect against the ravages of old age, other scientists are looking for compounds that turn back the clock entirely. Known as senolytic therapies, these drugs destroy the inflammation-producing zombie cells believed to be one of the causes of aging. A half-dozen companies are now involved in this effort, producing about a dozen drugs that obliterate zombie cells, delaying or alleviating everything from frailty and osteoporosis to cardiological dysfunction and neurological disorder.

Backed by investments from Jeff Bezos, the late Paul Allen, and Peter Thiel, Unity Biotechnology is one of the most interesting of these. They've developed a way to identify, then kill senolytic cells, or at least they've developed a way that works in mice. But it really works. Periodic treatments from midlife forward both extend lifespan by 35 percent and keep the mouse healthier along the way. Everything from lower energy levels to the development of cataracts and kidney dysfunction—all common symptoms of aging—are either avoided entirely or their onset is significantly delayed. With nearly a dozen drugs under development for nearly all the maladies of decrepitude, including a few that have completed Phase I human trials and are still moving forward, Unity remains one to watch in the anti-aging space.

Finally, there's Samumed, maybe the most watched of today's longevity companies. Backed by a $12 billion valuation, this San Diego–based biotech is focusing on the Wnt signaling pathways, which, like the name sounds, are one way the body sends messages. In this case, those messages govern a group of genes that both aid the growth of a developing fetus and seem to play a heavy role in aging. Errors in Wnt signaling have been directly linked to twenty different diseases, including cancer. This is also why these pathways have been a longtime target of almost every major pharmaceutical company. Samumed, though, may have cracked the code.

They've focused their efforts on one particular Wnt pathway that regulates the behavior of adult stem cells. Via this methodology, Samumed has developed nine different so-called "regenerative medicines." All are in the FDA's pipeline, including everything from hair-loss drugs to Alzheimer's drugs. Yet it's their success against arthritis and cancer that has garnered the most attention.

We'll start with arthritis, which afflicts 350 million people worldwide. We currently have no known treatment for the condition. But, in 2017, Samumed published the results of a small study on knee osteoarthritis. Sixty-one patients were in the trial, each receiving a single injection of a Wnt-rebalancing drug directly into their knee. All sixty-one saw improvement. When researchers measured the impact of the drug six months later, they found less pain and greater mobility, including an average of nearly two millimeters of new cartilage.

"The molecule stays [there] for about six months," explains Samumed CEO Osman Kibar, "during which it [stimulates] stem cells to grow new cartilage. And that new cartilage is that of a teenager. The key is that progenitor stem cells are there even when you're eighty years old, they just need to be properly signaled."

Yet this may only be the beginning.

"An injection of the same molecule into the spine of rats whose intervertebral disc has been destroyed regenerates a whole new disc," Kibar says. "If you look at the quality of the cells, the disc is younger and stronger."

Getting this to work in humans is a different story. Very few drugs make the leap from mice to men (or women), but other Samumed molecules developed to repair both rotator cuff and Achilles tendon injuries have already made it through Phase I trials, and their knee arthritis drug is now entering Phase III. There's still a lot more work to be done, but the upside could be a drug that provides decades of healthy mobility.

Arguably the more exciting arm of Samumed's research is their work in cancer—which is essentially stem cells gone haywire. By silencing the signaling pathway that leads to this frenzy, Samumed's drugs target—quite literally—every type of tumor. Most of these medicines are still in preclinical or Phase I safety and efficacy trials, yet, under compassionate use laws, Samumed has been able to give them to terminal patients. Here too the results have been remarkable.

In one small effort, a low-dose three-treatment protocol halted tumor growth in 80 percent of the study group. In another study involving pancreatic cancer, a longer protocol with this same compound managed to stave off this normally fatal disease. "All treatments had failed this woman," explains Kibar, talking about one patient's experience. "She weighed less than seventy pounds, and the doctors had sent her home [to die]. But now, after a year on our medicine, she's back to normal. She's traveling, dating, weighs one hundred and twenty pounds, just living her regular life. Of course, we're still in the early days for this compound, but it's a promising start to say the least."

The Bloody Fountain of Youth

In the early 2000s, a group of Stanford researchers went looking for the fountain of youth in an unusual place: the Dracula myth. Legends going back to the ancient Greeks, recaptured in the Roman poetry of Ovid, and revived again in Gothic vampire tales, talk of the rejuvenating effects of young blood. These Stanford researchers decided to test the theory—on mice.

Updating the gruesome, ancient technique of parabiosis, they linked together the circulatory system of a young mouse to the circulatory system of an old mouse, then pumped the former's blood through the latter's system. Results were visible to the naked eye. The young blood revived the older animal.

Upon closer inspection, the benefits went far beyond the visible. A variety of the old mouse's tissues and organs now had the characteristics of a far younger, healthier mouse. Follow-up studies both confirmed this finding and showed that the opposite was also true. Transfuse younger animals with blood from older ones and the clock spins forward, accelerating decrepitude, amplifying aging.

This work sparked a ton of interest. Within a decade, researchers had begun to tease apart why this youthful transformation was occurring. A team at Harvard got involved, discovering that young blood sparks the formation of new neurons in the brain and reverses age-related thickening of the walls of the heart. Finally, getting down to the root of the matter, the Harvard team also found one particular molecule, known as "growth differentiation factor 11," or GDF11, that appeared responsible for all, or at least some, of these benefits.

In a 2014 paper published in *Cell*, a different team of researchers showed that simply injecting GDF11 into mice increased strength, memory, and blood flow to the brain. Additional studies have extended these benefits, demonstrating that GDF11 can reduce age-related cardiac issues, accelerate muscle repair, improve exercise capacity, and amplify brain function.

All of this work has sparked entrepreneurial interest. The Harvard spinout Elevian, for example, led by entrepreneur Dr. Mark Allen and a quartet of the school's regenerative biology professors, is seeking longevity in GDF11 and similar age-retardant molecules. The Stanford spinout Alkahest, meanwhile, is searching for an optimized plasma cocktail as a treatment for Alzheimer's.

Wired called these kinds of efforts a "needle in a haystack approach," because blood plasma contains more than ten thousand different proteins. Really, it's a needle in a gold-stack approach, because identifying which proteins produce the young blood effect has led to a biological gold rush. Startups are hot on the trail, not to mention major pharmaceutical companies. In 2017, the National Institute on Aging committed $2.35 million in funds for scientists interested in the work. So yes, here in the twenty-first century, not only are flying cars and personal robots both suddenly real, but, well, so is Dracula.

For thousands of years, we've been searching for the location of the fountain of youth. But what all this work makes clear is that what we've actually been hunting is less a place than a time. The fountain of youth is a specific period in history, the point at which technologies converge on mortality. So while "Will we be able to live forever?" remains an unanswered question, turning a hundred years old into the new sixty—that is, significantly extending human lifespan—has changed from a question of "if" to a matter of "when."

CHAPTER ELEVEN

The Future of Insurance, Finance, and Real Estate

The Full Future

We've been working our way through the future, focusing on those sections of society with the most impact on our daily lives. So far we've covered transportation, healthcare, longevity, retail, advertising, entertainment, and education. In Part Three, we'll broaden our scope to the wider topics of energy, the environment, and government. But what about everything in between? In short, what about the future of the rest of your life?

That future will certainly be different from the past. Technological acceleration appears to be leaving no stone unturned. From architecture and artistry to aviation and accounting, dozens of industries will soon be transformed. Yet, in the final two chapters of Part II, we're going to turn our attention to four specific areas—insurance, finance, real estate, and food.

We've chosen this quartet for a number of reasons. For starters, three of these sectors—finance, insurance, and real estate—are top ten industries in America. In fact, when coupled with what we've already covered in Part II—excluding government and security, two topics we're saving for Part III—these sectors round out our look at the main ways Americans make a living. In this chapter we'll examine the first three. Then, in the next chapter, we'll close out Part II with a look at the future of food, because it's both foundationally important and wildly accelerating. Thus, in total, the sectors we'll review in this chapter are the front runners in the race to reinvent the rest of your life.

Let's dive in. . . .

Coffee, Risk, and the Origins of Insurance

In 1680, Edward Lloyd arrived in London. He was thirty-two years old and hunting for opportunity. He found one in coffee. Fueled by this then novel beverage, London's coffeehouse scene was exploding. Over three thousand java shops were already scattered throughout the city. Was the marketplace too crowded for another competitor? Lloyd didn't think so. In 1686, he launched his own establishment, Lloyd's Coffee House, on London's Tower Street.

At the time, London was driven by two economic engines: shipping and finance. Lloyd's was located at the epicenter of both, tucked into a tiny area between the Tower of London and Thames Street. Given this location, from the beginning the shop was popular with merchants, sailors, and shipowners.

Back then, coffeehouses earned patron loyalty by offering a mix of caffeinated beverages, the latest news and information, and the opportunity to participate in heated intellectual debate. Lloyd's offerings went further than most. To provide reliable and accurate shipping news to his customers, he established a network of correspondents in ports across Europe and published the news they gathered on tip sheets. Spending time at Lloyd's meant gaining access to detailed intelligence on ships, cargo, and foreign events.

This big data/big caffeine combo was a big winner. By 1691, business was booming and Lloyd needed to expand. He moved his operation to

16 Lombard Street, across from the Royal Exchange and into the heart of the merchants' quarter. This newer, much larger venue was tricked out with wall-to-wall blackboards and a central pulpit. The blackboards replaced the tip sheet. The central pulpit provided a place to announce maritime auction prices and real-time shipping news. And here, amidst black coffee and blackboards, Lloyd turned an idea first invented by the Babylonians into the foundation of the modern insurance industry.

Nearly four thousand years ago, the Babylonians developed a strategy for merchants sailing the Mediterranean. If a merchant took out a loan to fund a shipment, they would pay an additional sum in exchange for a guarantee: Should his goods be stolen or lost at sea, the lender would cancel the loan. In the fourth century BCE, loan rates differed according to the time of year. Prices were cheaper during the calm summer seas versus the more dangerous winter swells—that is, the Babylonians developed a risk-based pricing idea similar to the foundation of modern insurance.

Some two millennia later, this notion of risk-based, data-driven maritime insurance reached new heights in the confines of a certain London coffeehouse. The bankers who frequented Lloyd's were willing to collect premiums in exchange for taking shipping risks. They dubbed this process "underwriting," as bankers would literally write their names on the blackboard under the name of the ship and a list of the trip's details: its cargo, crew, weather, and destination.

Today, some 320 years later, this idea of "underwriting" has grown into a multitrillion-dollar insurance industry. Lloyd's humble coffeehouse evolved into the famed Lloyd's of London, which generated 33.6 billion pounds' worth of insurance premiums in 2017. Yet, driven by forces very similar to those that originally helped sculpt Lloyd's—an upsurge in information and collaboration—the insurance industry is again about to be completely transformed.

Three major changes are under way. First, by shifting the risk from the consumer to the service provider, entire categories of insurance are being eliminated. Next, crowdsurance is replacing traditional categories of health and life insurance. Finally, the rise of networks, sensors, and AI are rewriting the ways in which insurance is priced and sold, remaking the very nature of the industry.

But let's start with a simple question: If you're riding in an autonomous car as a service, and there is no driver, do you need insurance?

The Car That Doesn't Crash

Insurance is a game of averages. The industry's basic business model is assess risk and set premiums—or, covering this much risk will cost this much money. With a large enough number of customers and long enough stretches of time, this averages out to a profit for the underwriter. Car insurance premiums, for example, are currently calculated according to the age and history of the driver, traits of the car itself, and where the driver and that car live. Get enough drivers involved, stay in business long enough, and the result is massive profit. But what happens over the next decade, when autonomous vehicles take to the road and change every aspect of that calculation?

Right now, human error sits at the center of auto insurance. People—distractible, emotional, and sometimes irrational—are responsible for 90 percent of the 1.2 million traffic fatalities a year. Yet, without humans in the driver seat, 90 percent of that danger gets removed. For an insurance industry built on assessing risk, this alone is a huge change.

Now take it a step further. Today, we insure the stuff we own. But autonomous cars shift us from car-as-property to car-as-service, removing the need for consumer-facing auto insurance altogether. Which is why the accounting firm KPMG predicts the car insurance market could shrink by an astounding 60 percent by 2040.

This shrinking has already begun. Waymo automatically provides passengers with insurance every time they step inside one of their vehicles. And it's an assessment made with a confidence that comes from big data.

In 2018, Waymo vehicles had autonomously traveled some 10 million miles on public roads, and an additional 5 billion miles inside of a simulation. All of those trips were data-gathering missions, with the resulting information being used to train Waymo's AI. The big deal here is both safety and a nearly unassailable market advantage. All that data puts Waymo far ahead of the competition. It means that our autonomous car transition hasn't really begun and already traditional insurance companies are years behind the curve.

When we combine autonomous vehicle technology with smart traffic systems and sensor-embedded roads—two developments that have already begun rolling out—transit risks don't just plummet, they mutate. For instance, if the LIDAR sensor that's helping steer an autonomous car goes on the blink and causes an accident, who do you blame? Not the passenger. Maybe the carmaker. Maybe the LIDAR supplier. Or, who's fault is it if your Waymo loses its 5G connection and suddenly can't drive? Is it Alphabet, who owns the car; Verizon, who manages the connection; or OneWeb, who owns the satellite that provides that connection? What if an autonomous vehicle gets hacked or stolen?

While these questions remain to be answered, and certainly the scenarios they describe sound dangerous, it's helpful to remember that today, we routinely give testosterone-laden teenage boys control of two-ton vehicles. Not to mention the fact that roughly 1 million people are arrested for DUIs each year. In other words, just like the old tech, the new tech comes with trade-offs. But this time around, one of those trade-offs might be the end of auto insurance as we know it.

Crowdsurance

Before the arrival of exponential technology, size was the ultimate insurance advantage. Again, it comes down to averages or, more technically, what it takes to calculate an average in the first place.

Statistically accurate actuarial tables demand a ton of data. To gather that ton, you need an army of customers. To find those customers, you also need an army of salespeople. To analyze the data generated by these two armies, you need, no surprise, another army, of statisticians. Managing all these armies takes another army. And, up until now, this law of large armies guaranteed insurance was a game played by giants.

It was also a game of statistics. In both health and life insurance, the premiums of the healthy cover the costs of the unhealthy. But the healthy end up paying unnecessarily high premiums for this privilege, making them the consistent losers of this particular game.

So what happens when the ultra-healthy tire of this arrangement and decide to use social media to find other ultra-healthy folks, share data, and self-insure? All it takes is a few people raising their digital hand and saying, "Hey, look at my genes, see how much I exercise, check out my Oura ring data, my Apple Watch data. If anyone else is healthy like me, let's come together and do this."

In the insurance game, when the lowest risk clients opt out, the statistics stop working. With the ultra-healthy gone from the pool, the risk curve shifts dramatically. To cover costs, either everyone's rate increases or the insurance company goes bankrupt. But if everyone's rate goes up, then everyone goes elsewhere for insurance, and, once again, the insurance company goes bankrupt.

Which is exactly what's happening.

Introducing decentralized, peer-to-peer insurance, or what's become known as "crowdsurance." Crowdsurance eliminates the middleman. Instead of an insurance company, there's a technology stack—an app connected to a database connected to an AI-bot. The stack oversees a network made up of people who pay premiums and file claims which the network then approves. To put it differently, the stack removes three out of the four armies needed to create an insurance company. And the only remaining army? The customers—the ones left to decide what to do with all the money they just saved on health and life insurance.

Take the New York–based Lemonade, arguably the best funded of today's crowdsurance startups. Via an app, Lemonade brings together small groups of policyholders who pay premiums into a central "claim pool." Artificial intelligence does the rest. The entire experience is mobile, simple, and fast. Ninety seconds to get insured, three minutes to get a claim paid, and zero paperwork.

Adding more technology to this arrangement, companies like the Swiss firm Etherisc sell "bespoke insurance products" on the Ethereum blockchain. Because smart contracts remove the need for employees, paperwork, and all the rest, all sorts of new insurance products are being created. Etherisc's first offering is something not covered by traditional insurers: flight delays and cancellations. Individuals sign up via credit card, and if their plane is more than forty-five minutes late, they're paid instantly, automatically, and without the need for any paperwork. And they're only one example. The crowdsurance space is exploding. Brand new micro-insurance categories—boat-hull insurance, chihuahua insurance—are moving out of the planning stages and into the marketplace. To return to our historical terms, it's as if the sailors who frequented Lloyd's started making deals directly with the blackboards, and everybody else was just left to drink their coffee in silence.

Dynamic Risk

Founded in 1937, Progressive Insurance started out by trailblazing a niche nobody wanted: high-risk drivers. Then, true to their name, they maintained their edge via technology. Progressive was the first insurance company to have a website, the first to allow customers to purchase policies through that website, and the first to dress up that site with high-quality video content and voice-over-internet tools. By pioneering apps for policy purchasing and management, they were also first into mobile. All of these developments helped modernize insurance and made Progressive one of the more profitable corporations in America. But in 2004, they became first into a different category, and this decision was a little more than progressive. In an actuarial kind of way, it was downright revolutionary.

Not at the beginning.

At first, all Progressive did was ask customers in Minnesota to volunteer for a research program, "TripSense." Literally a black box, TripSense plugged into a car's diagnostic port and tracked three variables: mileage, speed, and travel times. When completed, volunteers shipped the box back, and Progressive sent them twenty-five bucks for their trouble.

In 2008, this pilot went mainstream. Renamed "Snapshot," this upgrade was designed to collect a single piece of information: vehicular speed at one-second intervals. Progressive then used that information to calculate two additional data points: miles driven and "hard-braking events," such as a driver slamming on the brakes to avoid hitting a cat. Why? Because Snapshot had morphed from a pilot project into a radical idea: price auto insurance according to driving habits rather than driving history.

The technical term for this is dynamic pricing. The paranoid term: Big Brother is always watching. Either way, Progressive helped replace the traditional auto insurance model with sensors. Everything from driving speed and braking habits to radio volume and the number of other cars on the road can impact your rate. Drivers now get insured based on car usage rates (the less you drive, the less you pay), good driving trends (you consistently stay within the speed limit), and low-risk driving times (your daily commute does not take place after midnight).

This same trend is creeping into home insurance. Pricing used to be based on the state of the home at policy purchase, but 30 percent of all home claims are from water damage that occurs long after the policy is sold. Now insurance companies get real-time metrics using in-pipe temperature sensors and in-wall water detectors, and home owners are notified about potential problems well before they occur.

Thanks to all the data from our wearables, this same shift will soon arrive in health insurance. Insurance companies will suddenly have the opportunity to prevent disease before it happens rather than swoop in post-op to clean up the mess. The upside will be cheaper insurance for healthier living, the downside is Big Brother. Do rates go up when you sneak a cigarette? Do they go down when you eat your vegetables?

The term McKinsey coined to describe this kind of AI-driven, sensor-laden insurance is "pay-as-you-live," morphing the traditional "detect and repair" role of an insurance company into "predict and prevent." Your rates fluctuate with your choices in an almost entirely automated process. By 2030, the number of humans required to process a claim will drop by 70 to 90 percent, while processing times will shrink from weeks to minutes. This points toward a future where insurance companies become front-line guardians of the health of society, and quite a change from the days of Lloyd, his coffee, and his blackboards.

Finance

Look at the skyline of São Paulo, Hong Kong, or New York. Check out the tallest buildings, those iron monsters. Who owns that high-flying real estate? Insurance companies and financial firms. Why? For the same reason that Willie Sutton allegedly said he robbed banks—"Because that's where they keep the money."

Since we've already covered insurance, we'll shift focus to the changes in banking and finance, where exponential technologies are steamrolling both industries, completely altering this business of money. We got a peak at this transformation earlier, when we decoded the dollars pouring into crowdfunding, ICOs, venture capital, and sovereign wealth funds. To understand what else is coming, let's start with a simple question: What, exactly, do we do with our money?

We store it, of course. Mostly in banks. We also move it around, sometimes transferring cash between companies, other times borrowing or lending among individuals. Next, we invest it, trying to use our money to grow more money. Finally, since the time coins were conch shells, we trade it for the stuff we want. Thanks to converging exponentials, each of these areas is being reimagined, with bits and bytes replacing dollars and cents, and neither economics nor the way we live our lives will ever be the same.

Good Money

Gunnar Lovelace grew up poor. He was raised by a single mother in an intentional community in California, and never forgot his family's struggle to meet basic needs, chief among them food and money. Lovelace went on to become a serial entrepreneur, and while his first three startups were in tech and fashion, he poured those profits into his fourth venture, Thrive Market, or his attempt to solve the struggle for food.

A purpose-driven venture, Thrive uses eco-friendly packaging, zero-waste shipping, and nontoxic ingredients to deliver high-quality organic food at lower prices directly to the door of over 9 million consumers. But Thrive only solves the first of Lovelace's problems, there's still the struggle for money to consider. This brings us to his fifth company, Good Money, which is using this same value-driven approach to take on the storage side of the traditional banking industry.

Right now, most of our money sits in banks, and mostly, we're abused for the privilege. On average, we pay $360 a year in banking fees. The larger banks, meanwhile, average $30 billion a year in overdraft charges alone. But where things go truly sideways is what the banks do with our money.

Banks get to invest our money, typically at significant profit, wherever they see fit. This often includes projects that don't align with customers' values. Wells Fargo, for example, lost a ton of business when they were outed as major backers of the controversial Dakota Pipeline. So while the bank is making bank, not only is your money not working for you, it might actually be working against you.

Good Money does the opposite, in a half-dozen different ways. Technically a mobile wallet, Good Money lives on your phone and holds both regular and crypto currencies. It can be used at any ATM, with zero annual fees, no ATM charges, and an interest rate a hundred times larger than most banks. Customers also become owners. Put money into Good Money and you get an equity share in return, while the company funnels 50 percent of their profits into impact investments and charitable donations.

With this strategy, Good Money is targeting people who prefer value-driven companies, and the 40 million Americans who have been driven out of the traditional banking system by overdraft fees and blackball lists. But the largest mobile market is a third category entirely, the unbanked, those without any place to store their money.

The issue is infrastructure, especially in poorer countries, where the cost of building and maintaining banks simply exceeds the value they can generate. The economics are upside down. Yet, while over two billion people in the world still lack bank accounts, nearly all of them have mobile phones. And this brings us to a Vodafone executive named Nick Hughes and one of the more stubborn of all international economic problems: microfinance.

An Unusual Proposition

At the 2002 World Summit for Sustainable Development, Vodafone's Nick Hughes gave a presentation about risk. His goal was a long shot: convince large corporations to help poorer nations by allocating research dollars for high risk/high reward ideas. An official from the UK's Department for International Development was in the audience. Afterward, he approached Hughes with an even more unusual proposition.

The Department had begun paying attention to mobile phone usage and noticed that in parts of Africa, people were treating mobile phone minutes as a quasi-currency, trading them for goods and services that would normally require cash. They saw potential here and, more importantly, had a million dollars to invest. If Vodafone agreed to match funds, the Department agreed to finance a pilot project.

Since borrowing money is one of the largest problems faced by the unbanked, their initial pilot project idea was microfinance. A microloan for a cow, motorbike, or sewing machine—that is, the startup costs for a small business—is often the start of the end of the cycle of poverty. By giving people a way to withdraw and repay their loan via cell phone minutes, the Department suspected they might jump-start entrepreneurship in countries that needed it most.

The result of this collaboration was M-Pesa, which was initially rolled out in Kenya in 2007. Without bank branches or ATMs, M-Pesa relies on an ancient technology: people. Individual agents sell cell-phone airtime in local markets, trading minutes for cash and vice versa. Customers load the airtime to their SIM card, then to their phone, turning the minutes into money, which can be sent to another via text message.

While microloans initially generated this plan, remittances turned it into a force. Being able to transfer money without banking fees allowed workers in the city to send money to relatives in the countryside—aka, remittances—which both saved them the 12 percent charged by the likes of Western Union and replaced the older method: giving a bus driver an envelope of cash and hoping for the best.

Eight months after launch, a million Kenyans were using M-Pesa. Today, it's nearly the entire country. According to research done at MIT, with nothing more than access to basic banking services, M-Pesa lifted 2 percent of Kenya's population—over two hundred thousand people—out of extreme poverty.

Nor is it just Kenya. M-Pesa now provides banking services to over 30 million people in ten different countries. In places rife with corruption, it's become a way for governments to protect against graft. In Afghanistan, it's how they now pay the army. In India, it's pensions. And it's no longer just M-Pesa offering such services.

In Bangladesh, bKash now serves over 23 million users; in China, Alipay serves just shy of a billion. And like Good Money, Alipay has become a force for social good. More than 500 million customers play "Ant Forest," earning points for making environmentally friendly decisions in their daily lives. These points are then redeemed for real trees planted in the real world. It's become something of a national obsession. To date, more than 1 million trees have been planted.

Importantly, these developments upend the traditional arc of technology. Typically, cutting-edge ideas pioneered in Silicon Valley move from West Coast introduction to East Coast adoption to European investigation and finally, eventually, into the rest of the world. But this process has inverted, with developing world innovation becoming developed world disruption.

And more disruption is coming.

Banks occupy a rare spot in the economic ecosystem: All the infrastructure that money flows through belongs to them. As the trusted central repository, whenever anyone wants to move money around—lend it, transfer it, even give it away—banks can insert themselves into that process. Or, at least, until blockchain came along.

With blockchain, since trust is built into the system, the system is no longer necessary. Take a stock trade. Right now, to execute that trade, there's a buyer, a seller, a series of banks that hold their money, the stock exchange itself, clearinghouses, etc.—roughly, ten different intermediaries. Blockchain removes everyone but the buyer and seller. The technology does the rest.

In an attempt to hold on to their thinning slice of the pie, every major bank is rushing into blockchain. Yet arguably moving faster are the thousands of entrepreneurs using blockchain to disrupt these same banks. Consider R3 and Ripple, two examples of developing world disruption impacting developed world businesses. In both cases, these companies are using blockchain to replace the SWIFT network, the standard protocol overseeing international banking transactions.

This inverted flow of disruption isn't going to end any time soon. Over the next decade, 4 billion people, the rising billions, will gain access to the internet. Since all of them will need basic banking services, the opportunity is massive. But thanks to converging technology, the downstream result of Nick Hughes's long shot is that nearly everyone but banks will end up capitalizing on it.

The AI Invasion

"Fintech" is the term for the convergence of technology and financial services. First colonized by networks and apps, it was then radicalized by AI and blockchain, and now serves as a global wealth redistribution mechanism. Think Robin Hood with a smartphone, taking cash away from banks and putting it back into the hands of customers.

Wherever large amounts of customer frustration encounter large piles of money, opportunity lurks. This gave rise to a company called TransferWise. By matching customers who have, say, pesos they want to turn into dollars with customers who want to change dollars into pesos, TransferWise is using a modified dating app to take on the entire foreign currency exchange market. In fact, because it's easier to match people looking to exchange currency than it is to match people looking to date, the company reached a billion-dollar valuation in under five years.

Built on networks and apps, TransferWise is also an example of fintech's colonization wave. The radicalized wave arose when AI entered the picture. Consider the age-old practice of "Buddy, can I borrow a dollar," otherwise known as peer-to-peer lending. Traditionally, this is a high-risk practice—which is to say, Buddy rarely gets his dollar back. This problem only gets worse with scale. As villages turned into towns, towns expanded into cities, and cities began to sprawl, neighborly trust broke down. That's where banks came into play—they added trust back into the lending equation.

But who needs trust when there's data?

With AI, huge groups of people can come together, share financial information, and pool risk, becoming the peer-to-peer market now known as "crowdlending." Prosper, Funding Circle, and LendingTree are three examples in a market expected to grow from $26.16 billion in 2015 to $897.85 billion by 2024.

A different example is the Smart Finance Group. Created in 2013 to serve China's massive unbanked and underbanked population, Smart

Finance uses an AI to comb a user's personal data—social media data, smartphone data, educational and employment history, etc.—to generate a reliable credit score nearly instantly. With this method, they can approve a peer-to-peer loan in under eight seconds, including microloans to the unbanked. And the results speak for themselves. Roughly 1.5 to 2 million loans are taken out every month via Smart Finance.

AI is also making an impact on investing. Traditionally, this game was played by the wealthy because it's a game of data. Financial advisors had the best data, but you needed to be wealthy enough to afford a financial advisor to access it. And advisors are picky. Since it can take more time to manage small investors than large investors, many wealth managers have investment minimums in the range of hundreds of thousands of dollars.

But AI has leveled the playing field. Today, robo-advisors like Wealthfront and Betterment are bringing wealth management to the masses. Via an app, clients answer a series of initial questions about risk tolerances, investment goals, and retirement aims, and then algorithms take over.

Actually, algorithms have already taken over. Every day, roughly 60 percent of all market trades are made by computer. When the market turns volatile, this can climb to as high as 90 percent. All robo-advisors have done is make the process available to the customer, and save the customer money as a result.

With no humans in the chain, fees are slashed. Instead of the typical 2 percent of profits charged by a wealth manager, most robo-advisors take around .25 percent. Investors are responding. As of January 2019, Wealthfront had $11 billion under management, while Betterment was at $14 billion. While robo-advisors still account for only roughly 1 percent of total U.S. investment, *Business Insider Intelligence* estimates that number will climb to $4.6 trillion by 2022.

Finally we come to our last category, using money to pay for things. But we already know this story. When was the last time you dropped coins into a toll booth? Or paid cash for a cab ride? In fact, Uber and

Lyft allow us to get around a city without a wallet. Couple cashier-less stores like Amazon Go with services like Uber Eats and these wallet-less ways are about to become the new normal.

Denmark stopped printing money in 2017. The year prior, in an attempt to expand mobile banking and demonetize the country's gray-market economy, India recalled 86 percent of its cash. Vietnam wants retail to be 90 percent cashless by 2020. Sweden, where over 80 percent of all transactions are digital, is almost there.

Economists often point out that two of the main factors that drive economic growth are the availability of money—the stockpiles we can draw upon—and the velocity of money, or the speed and ease with which we can move that money around. Both of these factors are being amplified by exponential technologies. The result is acceleration, a profound shift in who owns the real estate that dominates our skylines. In fact, in turning to real estate itself, we will see that those changing skylines are only the start of this transformation.

Real Estate

Our story begins in the Great Recession of 2008, where the reckless behavior of two familiar players—big banks and big insurance—plunged the country into chaos. Real estate was especially hard hit. The bottom fell out of the market, and kept on falling. The housing crisis ensued, and things went from bad to worse to downright miserable. Which is when Glenn Sanford had what really should have been a stupid idea: start a real estate company.

But it was a company built for the times. In the face of rising overhead costs and decimated revenues, Sanford decided not to do what real estate agents had always done: open an office. Ditching the brick-and-mortar model entirely, he created eXp Realty, the first cloud-based national real estate brokerage.

Devoid of *real* estate, Sanford built a fully immersive mega-campus using a virtual world platform called VirBELA (which eXp now owns).

Today, eXp Realty's campus is home to sixteen thousand agents from all fifty US states, three Canadian provinces, and four hundred major real estate markets—all supported without a single office.

Instead of coming into work, agents and managers stay home. Using either a VR headset or their laptop, they gather virtually, on a campus replete with a lobby, library, theaters, meeting rooms, and a sports field. Sanford, meanwhile, attributes at least $100 million of his company's $650 million market cap to money saved via reduced infrastructure and overhead.

While Sanford's reinvention has had significant consequences for the real estate industry, it was merely the result of a trio of convergences: computation, networks, and VR. Pile on what's coming: AI, 3-D printing, autonomous cars, aerial taxis, and floating cities, and everything changes fast. And that includes the one component of the industry that Sanford left intact: the real estate broker.

Say Goodbye to Your Broker

Most people, if they're lucky, get to buy a home. Typically, it's once in a lifetime. Often, it's the largest purchasing decision they'll ever make, the biggest check they'll ever cut, and for many, the most terror they'll ever feel. We're about to tell you a story about how AI is changing this process, but before we get there, it's helpful to remember this is really a story about nervous people making hard decisions in the real world.

Truthfully, AI has been helping people make hard real estate decisions for a little while now. Zillow, Trulia, Move, Redfin, and many other companies have invested millions in the technology. Already, property searches, evaluation, consulting, and management are easier, faster, and more accurate than ever before. Investors now have analytical capabilities far beyond any human, blending established real estate variables such as rent, occupancy rates, and local school data, with novel inputs: Web clickstream data, satellite imaging, geolocation tracking, etc. And like the AI arms race in other industries, in real estate too, the companies with the best data will end up dominating the market.

If AI is already doing all this on the research side of real estate, why not let it handle brokering as well? Or, more specifically, why not let the convergence of AI, VR, and sensors become your broker?

In a future where always-on preference tracking is a standard feature of any personal AI, why hire a stranger to sell you real estate? Your AI already knows you, via your likes and dislikes, possibly better than you know yourself. We're not long from the day when most of your real estate search—home, apartment, office, whatever—will be conducted from your couch, via a VR headset, with the help of your personal AI. It'll be like talking to Siri if Siri were an interior designer: "Industrial modern loft, concrete floors, close to a Whole Foods," etc. Your real estate AI will offer up selections that fit your criteria, while the VR headset makes taking a tour a 24-7 proposition.

As a seller, this means potential buyers can come from two miles to two continents away. As a buyer, each immersive VR tour is an AI learning experience. Advanced eye-tracking software follows your gaze, while voice recognition algorithms listen for pleasure or aversion in your vocal tone. Both add to your list of likes and dislikes, with the AI recommending better and better properties as your search goes along.

Curious what this living room might look like with a fresh coat of blue paint? No problem. VR programs can modify environments instantly, changing countless variables, from flooring to wallpaper to the sun's orientation. Keen to input your own furniture into a VR-rendered home? Advanced AI could one day compile all your existing furniture, art, and books, then impose them on any virtual space, providing certainty in a place where we've never had much before. In essence, AI-driven, VR real estate platforms will allow you to explore any property on the market, do any remodel you'd like, then figure out if the home you dreamed up is really the home of your dreams.

Reinventing the City

"Location, location, location" is the real estate mantra, but there's an important addition to this story: "Proximity, proximity, proximity."

The value of your home is partially measured in its proximity to a half-dozen locations: a central shopping district, the best schools, your place of work, favorite restaurants, the homes of your closest friends, etc. But over the next decade, with the transportation transformation that's coming, we're changing the relationship between here and there. So what happens when driverless cars, flying cars, and the Hyperloop make proximity a possibility for all?

If Las Vegas to Los Angeles becomes a half-hour Hyperloop commute; if upstate Vermont to middle-of-Boston becomes a flying car ride away; if remote Virginia into Washington, DC, is an hour-long nap in an autonomous Uber, why not buy twice-the-house for half the price in outlying areas? Travel time can now be repurposed any way you like—sleep, meditation, conversation, take your pick. When previous geographically undesirable locations become easy to reach, proximity itself becomes democratized.

The point is that our definition of "prime real estate" is going to start to shift over the next decade. But this isn't just about altering the relationship between here and there—it's also about making more "here."

Enter: floating cities.

Floating cities are a proposed solution to a trilogy of modern problems: rising seas, skyrocketing populations, and threatened ecosystems. In the five hundred coastal cities now threatened by global warming, they could offer what we need most: flood-proof, tsunami-proof, and hurricane-proof living. Plus, close to 40 percent of humanity already lives near the ocean, so, if successful, floating cities will create an abundance of prime real estate in places where none has existed before.

Earlier versions of this notion met considerable resistance, but in 2019, in the face of mounting climate change dangers, the United

Nations decided the technology was worth a second look. One idea they've been considering is Oceanix City, a zero-waste, energy-positive design by Tahitian entrepreneurs Marc Collins and Itai Madamombe. It consists of a series of hexagonal islands arranged in a circle, with each of these self-sustaining 4.5-acre platforms able to house three hundred people, with the complete 75-acre property capable of sustaining up to ten thousand.

A second design by the San Francisco–based Seasteading Institute is being tested in the waters off French Polynesia. Known as the Floating Island Project, the idea here is less a floating city and more a test platform for the designs of future floating cities. With a hundred acres of beachfront property and a special economic zone for inhabitants, this project is aiming to have a dozen structures erected by 2021.

In both locations, sustainability is key. Water capture technologies provide drinking water; an array of greenhouses, vertical farms, and fish farms supply the food; and sunlight, wind power, and wave energy power the whole lot. Electric boats, or soon, autonomous flying cars, will zoom inhabitants to work on the mainland. Or, more importantly, maybe not. With drone delivery for supplies, and avatar commuting for business meetings, you may never have to leave the island.

Exponential technology is dematerializing, demonetizing, and democratizing nearly every aspect of real estate. Corporate infrastructure has gone virtual; brokers aren't far behind. Location and proximity, the twin pillars of the industry, are next, arguably demonetized before this decade's done. Today's scarcity of bespoke addresses for the lucky few will become tomorrow's prime real estate at affordable prices for the average many.

When you couple the changes in real estate to the remaking of finance and insurance, we're seeing a drastic shift in both the skylines of cities and the business of business. By making these processes faster and cheaper, cutting out the middleman (or woman), and making these opportunities available to all, we're reimagining three of the largest wealth creation engines in history. Which is, once again, another reason that the future will be faster than you think.

CHAPTER TWELVE

The Future of Food

It's 2030 and you're hungry. Recently, you've been checking out the latest craze: hybrid cuisines. Your AI has a few suggestions based on everything from your historical preferences to your nutritional needs and your calendar. Tomorrow, you're going surfing, so packing on a few extra calories beforehand is a good idea. Tonight's choice: Asian-Yiddish fusion. Since you've never tried this hybrid before, better safe than sorry. You let your AI select the actual dishes.

Eight minutes later, an Amazon drone arrives with two bags of raw ingredients. You unload these into seven bays attached to your 3-D food printer. There's a vegetable you've never seen before. Scanning a small code brings up a blockchain-backed food identity tracking app. Turns out, it's a new kind of squash, originally from Vietnam, now sourced at a vertical farm just down the road.

The remaining ingredients get passed over to your robo-chef, really nothing more than a pair of articulated arms with a touch screen interface. But no need to touch that screen, as the recipes were uploaded back when you first ordered dinner. Since the system is fully automated, no need to hang around either.

Heading out of the kitchen, you notice the robot arm slicing fresh tuna with a smooth motion. With 20 different motors, 24 joints, and

129 sensors, this kitchen-bot can mimic the movements of a human hand and arm. In fact, this particular bot was trained using machine learning from videos of top chefs cooking at five-star restaurants.

Even better, since you've selected Moo-Less Meats as your lab-grown provider, you know this tuna wasn't caught using bottom-trawling, dynamite fishing, or any other ecological nightmare. Instead, it was grown from stem cells, with no animals or environments harmed along the way. Finally, since the entire process has been both automated and customized, there is zero food waste. You eat everything on your plate and then, because it's been 3-D printed out of chocolate, you eat the plate.

In 2020, just about everything in this story is already here. Sure, it's not in your kitchen, but it will be soon. Before we consider the modifications to your apartment this might require, let's start our investigation into the future of food, where the food under investigation actually starts—in the nuclear heart of the star we call the Sun.

The Inefficiency of Food

The tale of food is a tale of waste. Inefficiency is built into every step of the story. Consider the origins of the meal on your plate. "All animals eat plants or eat animals that eat plants," author Richard Manning wrote in an essay for *Harpers*. "This is the food chain, and pulling it is the unique ability of plants to turn sunlight into stored energy in the form of carbohydrates, the basic fuel of all animals. Solar-powered photosynthesis is the only way to make this fuel. There is no alternative to plant energy, just as there is no alternative to oxygen."

The food on our plates begins its journey some 93 million miles away, in the solar portion of solar-powered photosynthesis. Even though millions and millions of metric tons of hydrogen are fused every second, less than one-billionth of that energy actually reaches the Earth. And, of the total that does hit the planet's surface, less than 1 percent is actually used for photosynthesis.

But the squandering doesn't end there. Once our food is grown, it still needs to be transported. Nothing about this process is easy on the environment. At holiday dinners, there is a pretty good chance the food on the table logged more travel miles than the family around the table. The average American meal travels fifteen hundred to twenty-five hundred miles to get to your plate. Potatoes from Iowa, wine from France, beef from Argentina—you get the energy-intensive picture.

Even more energy is squandered on the back end of that same meal. While one out of eight Americans struggles to put food on the table, 40 percent of the food in the United States is never eaten. Either it rots in the field or ends up in a landfill. In fact, according to the National Resources Defense Council, if we could "rescue" just 15 percent of that food, we'd be able to feed 25 million of the 42 million Americans now facing food insecurity.

Help is on the way, as every step in our food chain is now being completely transformed. At the front end of this process, researchers are beginning to figure out how to ramp up a plant's ability to turn sunlight into food. Tobacco is the lab rat of plant biology. By improving a plant's ability to use sunlight to make sugar, researchers at UCLA got a 14 to 20 percent boost in tobacco crop yield. The Bill Gates–backed RIPE Project at the University of Illinois has matched and improved those numbers. Studies run at the University of Essex took this farther. By increasing the levels of a protein involved in photorespiration, they improved tobacco yields by 27 to 47 percent. The UN estimates that we need to double overall crop production by 2050 to feed a population estimated at over 9 billion. What all this work shows is that improving photosynthesis can get us more than halfway there.

These upgrades may take a while to move from lab to table, but companies are already preparing for the next step in this process: transportation. Not only are our vehicles becoming more energy-wise, our food is becoming more durable.

The Santa Barbara–based Apeel Sciences is using biomimicry and materials science to take on the problem of food waste. Nature, it turns out, endows fruits and vegetables with a natural anti-spoilage mechanism: the peel. Technically known as "cutin," this outer layer of a plant's epidermis is a waxy skin of fatty acids meant to trap moisture in. Apeel has found a way to use completely natural, plant-derived materials to make cutin in the lab, where it can be sprayed onto food (or the food can be dipped in it). It's odorless, tasteless, colorless, and food coated with it is still considered organic. Avocados, protected in this manner, take 60 percent longer to soften and are already available at most major American grocery stores.

Protecting against spoilage helps preserve our food for longer, but it doesn't completely solve the transportation question. So companies have started to bypass this step altogether. In order to make moving food from farm to table more efficient, they're moving the farm. It's an idea known as "vertical farming," or growing food inside of skyscrapers rather than outside in fields. Since over 70 percent of humanity will live in cities by 2025, driving farm-grown vegetables an average of two thousand miles to reach our urban plates is not only wasteful, it's unhealthy. Every second a plant is out of the ground, its nutritional value decreases. If it takes two weeks for food to make it to your table—a not uncommon travel time—it can reduce the nutritional value by as much as 45 percent. With vertical farms, locally grown means locally grown. IKEA, for example, now feeds customers in their stores with food grown vertically in their stores.

Beyond eliminating travel time, vertical farms solve a number of other problems. Since they are completely closed environments, the need for pesticides also disappears. As does the need for water. Relying on hydroponics and aeroponics, vertical farms allow us to grow crops with 90 percent less water than traditional agriculture—which is critical on our increasingly thirsty planet.

Vertical farming has been progressing at an incredible rate. Back in 2012, when we first wrote about the concept in *Abundance*, there were merely a handful of pilot projects in existence. Today, there's an industry.

The largest player around is the Bay Area–based Plenty Unlimited Inc. With over $200 million in funding, Plenty is taking a smart tech approach to indoor agriculture. Plants grow on twenty-foot-high towers, monitored by tens of thousands of cameras and sensors, and optimized by big data machine learning. This allows them to pack forty plants into the space previously occupied by one. It also produces yields 350 times greater than outdoor farmland, and uses less than 1 percent as much water. And rather than bespoke veggies for the wealthy few, Plenty's processes allow them to knock 20 to 35 percent off the costs of traditional grocery stores. To date, Plenty has their flagship operation in South San Francisco; a one-hundred-thousand-square-foot farm in Kent, Washington; an indoor farm in the United Arab Emirates; and they have started construction on over three hundred farms in China.

On the other side of the US, in a repurposed seventy-thousand-square-foot factory in Newark, New Jersey, a company called AeroFarms has figured out how to grow over 2 million pounds of leafy greens a year with neither sun nor soil. Instead, inside their facility, rows of AI-controlled LEDs provide the exact wavelength of light each plant needs to thrive. Using aeroponics, nutrients are misted directly onto the plants' roots, so no soil is required. Rather, plants are suspended in a growth mesh fabric made from recycled water bottles. And here too, sensors, cameras, and machine learning govern the entire process.

None of these farms are yet big enough to impact our global food problem, but the exponentials are conspiring in their favor. Agriculture, as Plenty CEO Matt Barnard recently told reporters, is seriously benefiting from convergence. "Just like Google benefited from the simultaneous combination of improved technology, better algorithms and masses of data, we are seeing the same [in vertical farming]."

We're also seeing robots enter the picture. Right now, 50 to 80 percent of a vertical farm's cost is human labor, but the Silicon Valley–based Iron Ox has designed a thousand-pound robot that can tote around eight-hundred-pound growing containers. Or, as *Engadget* recently wrote: "Old MacDonald was a droid."

In short, not only is farming getting taller, it's getting stronger. And smarter. And most importantly: far more efficient.

The Inefficiency of Growing a Cow

Scary fact: By 2050, to feed a population of 9 billion, the world will need 70 percent more food than it did in 2009. A lot of that food will be meat. By 2050, primarily thanks to the modernization of China and India, global meat consumption is expected to increase by 76 percent. This is problematic, to say the least.

Today, 50 percent of all habitable land on Earth is used for agriculture, with 80 percent of that land reserved for livestock. A quarter of the planet's available landmass is currently used to keep 20 billion chickens, 1.5 billion cattle, and 1 billion sheep alive—that is, until we can kill and eat them. The suffering quotient is through the roof. As is the waste. One out of eight Americans will go to bed hungry tonight, yet farm animals consume 30 percent of the world's food crops.

Worse is the water involved. Meat production accounts for 70 percent of global water use. Compared to fifteen hundred liters required to produce a kilogram of wheat, it takes fifteen thousand liters to produce a kilogram of beef, meaning there's enough water in an adult steer to float a US Navy destroyer.

Meat is also responsible for 14.5 percent of all greenhouse gases and a considerable portion of our deforestation problem. In fact, we're in the midst of one of the largest mass extinctions in history—and more on this in Chapter Thirteen—and loss of land for agricultural production is currently the largest driver of that extinction.

The core of the issue is the same: inefficiency.

Today, we must raise an entire cow to produce a single steak. We also need to deal with all the waste and the greenhouse gases that cow produces along the way, and dispose of the animal's carcass on the back end. Yet, as advances in biotechnology have begun converging with advances in agrotechnology, we can now bypass this entire process, growing that same steak from a single stem cell—no cow required.

Here's the cultured meat formula. Take a few stem cells from a live animal, typically via a biopsy so the animal isn't harmed. Feed these cells a nutrient-rich solution. Power the whole process in bioreactors. Give the industry a few years to mature and the technology a few years to shed costs and, finally, we can produce an infinite number of steaks to feed an increasingly carnivorous population.

Or, at least, that's the goal, but we still have a few hurdles to clear. As of today, the nutrient-rich solution remains animal-based and riotously expensive. If one of the main aims is cruelty-free meat, the solution will have to be entirely plant-derived, which scientists and companies are still working on. Due to our inability to deliver this solution with precise timing and locational accuracy, we've had more success mimicking "softer" meats—like raw hamburger or chorizo—than growing steak itself. Lastly, in a collaboration between the food and energy industries, researchers are still working on ways to power the whole system. Eventually, bioreactors will require less power and/or could run completely off renewables, but we're not there yet.

Still, the environmental benefits are considerable. Cultured meat uses 99 percent less land, 82 to 96 percent less water, and produces 78 to 96 percent less greenhouse gases. Energy use drops somewhere between 7 and 45 percent depending on the meat involved (traditional chicken ranching is much more energy intensive than traditional beef ranching). By freeing up a quarter of our landmass, we can reforest, providing the habitat required to halt the biodiversity crisis and the carbon sinks needed to slow global warming. And while that's a haze of numbers to consider, what they point to is amazing: an ethical and environmental solution to world hunger.

Cultured meat is also a healthier solution. Since we're growing steak from stem cells, we can increase helpful proteins, reduce saturated fat, and even add vitamins. The meat requires no antibiotics and, given the danger of diseases like mad cow, is actually safer for humans. By turning to cultured meat, we're lowering the global disease burden and, since 70 percent of emerging diseases come from livestock, decreasing our risk of pandemic. Plus, tests with both consumers and chefs have shown that taste is no longer a factor.

Currently, doing this at scale remains expensive. Back in 2013, the first cultured burger cost $330,000. By 2018, Memphis Meats had gotten that down to $2,400 a pound, while Aleph Farms has steak down to around $50 a pound. Yet again, exponentials are on their side: Memphis believes accelerating technology will bring that burger cost down to around $5 within a few years, and in higher end Asian restaurants, lab-grown chicken is already on the menu.

As it progresses, cultured meat has the potential to become far more cost-effective than conventional meat. Production is mostly automated, without much need for land or labor. It also takes a few years to grow a cow, but only a few weeks to grow a cow's worth of steak in the lab. And it's more than steak. The meats in development range from pork sausage and chicken nuggets to foie gras and filet mignon—it all depends on which cells you start out with. In late 2018, Just Inc. announced a partnership with Japanese Wagyu beef producer Toriyama to develop meat from the stem cells of what has long been the rarest and most expensive steak on Earth.

What's true for meat is also true for milk. Perfect Day Foods, a Berkeley, California–based company, has figured out how to make cheese without need of cows. By blending gene sequencing with 3-D printing and fermentation science, they've created a line of animal-free dairy products.

Add this all together and we see a very different future of food. In a few years, humans will become the first animals that get their protein from other animals without any animals being harmed along the way. Slaughterhouses will become a ghost story we tell our grandchildren. And a planet that is already strained under the weight of nearly 8 billion souls will have a fighting chance when our numbers top 9 billion.

CHAPTER THIRTEEN

Threats and Solutions

Let's widen our scope. Up to now, this book has had two primary goals. In Part One, we examined the forces of acceleration, seeing how converging technology is unleashing waves of change historically unrivaled in their capacity for disruption. In Part Two, we tracked those waves through society, paying particular attention to their impact on our day-to-day lives. In both cases, we kept the scope of our examination limited to the next ten years.

In Part Three, we're broadening our horizons in two key directions. In this chapter, we'll focus on disruptions to our disruption—that is, a series of environmental, economic, and existential risks that threaten the progress we've made. Obviously, each of these categories is thick enough to fill a textbook. Our goal isn't the exacting details. Rather, we want to outline the problem, then examine the ways in which converging technologies might provide solutions.

In our next chapter, we're going to take the longer view, expanding our focus from the coming decade to the century ahead. We'll explore five great (mostly) tech-driven migrations that are beginning to unfold. We'll unpack economic relocations, climate-change migrations, virtual worlds explorations, outer space colonization, and hive-mind collaborations—or the quintet of mass movements that will reshape the demographics of the globe and the nature of society over the next hundred years.

We'll begin by turning our attention to the ongoing water crisis, then expand into climate change and species die-off, before shifting focus to technological unemployment, rogue AIs, and other threats that go bump in the exponential night.

Water Woes

In 2018, the United Nations Intergovernmental Panel on Climate Change released their "Special Report on Global Warming," which reached a stark conclusion: We humans have broken the planet. By falling in love with industrial technology and failing to address the environmental devastation it produced, humanity has put the Earth—what Carl Sagan once called, "the only home we have ever known"—on a crash course with disaster. According to the world's top climate scientists, we have just twelve years to fix this problem. Either limit global warming to 1.5 degrees or face catastrophic consequences.

A few months later, the World Economic Forum seconded that notion, releasing their Global Risks Report, the latest edition of a periodic issuance designed to highlight the top five threats humanity will face over the next decade. Traditionally, the WEF's concerns are economic—oil crises, financial crashes, that sort of thing. But 2018 marked the first time that fiscal fears didn't make the top five. Instead, today's biggest dangers are all ecological in nature: water crises, biodiversity loss, extreme weather, climate change, and pollution.

Over the next few sections, we'll examine how technology is helping us tackle the WEF's top five concerns, but this doesn't happen automatically. Ours is not a techno-utopian argument. Solving our planet's ecological woes requires technology, for certain, but it also demands one of the largest cooperative efforts in history. If we can learn to work together like never before, we like our chances. But in light of these recent reports, sooner rather than later.

And this brings us to Dean Kamen.

Dean Kamen is a kind of geek superhero, a nerd Batman in a denim work shirt. For starters, he lives in a secret lair—an island fortress complete with hidden rooms, helicopter launchpads, and after peacefully seceding from the United States, its own constitution. His résumé includes over 440 different patents, including insulin pumps, robotic prosthetics, and all-terrain wheelchairs. Because so many of his inventions have had such an impact, in 2000, President Bill Clinton awarded Kamen the highest honor awarded to inventors, the National Medal of Technology.

In *Abundance*, we recounted the story of Kamen's "Slingshot," an innovation named after the weapon David used to defeat Goliath, yet designed to fell a different giant: the threat of water scarcity. Today, 900 million people lack access to clean drinking water. Waterborne diseases are the number one killer on Earth, claiming 3.4 million lives a year, most of them children. Climate change, our rapidly ballooning population, and consistently poor resource management aren't helping matters. By 2025, according to the UN, half the globe will be water stressed.

To turn this tide, Kamen designed the Slingshot, a vapor compression distillation system powered by a Stirling engine—or, a water purifier the size of a mini-fridge capable of running off any combustible fuel, including dried cow dung. Using less electricity than required to power a hair dryer, the Slingshot can purify water from any source: polluted groundwater, saltwater, sewage, urine, take your pick. One machine provides clean drinking water for three hundred people a day; a hundred thousand machines—now that's the kind of cooperative effort we're talking about.

This is also why we want to return to Kamen's story, picking up in 2012, where we left off. Back then, the Slingshot had just completed a round of beta testing, successfully providing a couple months of clean drinking water to a number of remote African villages. Simultaneously, Kamen had just made a handshake deal with Coca-Cola. The inventor agreed to build the soft drink behemoth a better soda fountain, and in return, Coke agreed to use their global distribution network to get the Slingshot into water-starved countries.

Both kept their word. Kamen helped design the "Freestyle Fountain Beverage Dispenser," which uses "micro-dose technology" to mix over 150 different beverages on demand (talk about choice paralysis). Coca-Cola, meanwhile, teamed up with ten other international organizations and began distributing the Slingshot in 2013, a core feature of their "Ekocenter" kiosks. Part general store and part community center, Ekocenters are solar-powered shipping containers that provide remote, low-income communities with safe drinking water, internet access, nonperishables (like mosquito repellant), first-aid supplies, and, of course, Coca-Cola products for sale. By 2017, there were 150 Ekocenters operating in eight countries, most of them run sustainably, by local female entrepreneurs, distributing 78.1 million liters of safe drinking water a year—not bad for a handshake deal.

Yet the Slingshot is not the only deal in town.

Technology has begun converging on our water woes, with thousands of players working on an enormous range of approaches. There are high-tech nanotechnology-infused desalination plants and medium-tech solar-powered groundwater pumps and low-tech fog capture methods. To offer another example, Kamen's Slingshot has competition from the Bill Gates–backed Omni Processor, which turns human feces into potable drinking water while simultaneously producing electricity for power and ash for fertilizer.

There's also the California-based company, Skysource, winner of the $1.5 million Water Abundance XPRIZE, whose technology extracts two thousand liters of water a day from the atmosphere—or enough for two hundred people. It does this with renewable energy and at a cost of no more than 2 cents a liter. As the daily water needs for a planet of 7 billion are between 350 and 400 million gallons a day, using technologies like Skysource to tap the more than 12 quadrillion gallons contained in the atmosphere at any one time might be the only way to quench that thirst.

Or consider the "smart grid for water," which is what happens when exponential technologies converge on the farm. The smart grid allows for everything from precise soil monitoring and crop watering to the early detection of insects and disease. Estimates vary, but most studies find the smart grid capable of saving us trillions of gallons a year—which is the point. We're not lacking in technological know-how. We are water-wise, but execution dumb, attacking a biosphere-wide problem with a piecemeal approach.

Yet this is also the typical developmental curve for exponentials. Water technologies are moving out of the deceptive and into the disruptive phase, stitching these piecemeal efforts together into the global solutions we actually need. And one reason we can say this with confidence is that water technologies appear to be about five years behind energy technologies, which—as we'll soon see—are scaling up into a worldwide force for tackling the next problem we'll consider: global warming.

Climate Change for Optimists

Forty billion tons of CO_2—that's the cost of burning fossil fuels. Every year, we dump 40 billion tons of carbon dioxide into the atmosphere. How to wrap your mind around that amount? In 2017, *Scientific American* journalist Caleb Scharf tried to find a comparison and lit upon forest fires.

Trees store carbon. If you burn an acre of coniferous forest, it releases 4.81 tons of carbon. Therefore, to release 40 billion tons of carbon requires burning 10 billion acres of forest per year, or the equivalent of 42 million square kilometers. Unfortunately, explained Scharf, "the entire continent of Africa is a mere 30 million square kilometers. So [Africa], plus another third, on fire, each year."

These emissions, the detritus of incinerated coal, oil, and natural gas, are the principle driver of global warming. In fact, according to the Carbon Majors Database, since 1988, 71 percent of greenhouse gas emissions can be backtracked to just a hundred fossil fuel companies. For this reason switching to clean energy tops pretty much every list of what we can do to stop climate change. And most experts agree, that switch has three major parts: energy generation, energy storage, and green transportation. So here, in our examination of the solutions to the biggest threats we now face, we'll start with energy generation—where the news is good.

Wind and solar power have been riding exponential growth curves for decades, dropping in price and rising in performance on an incredibly consistent basis. For comparison purposes, coal, which has long been the cheapest energy available, now costs about 6 cents a kilowatt-hour (kwh). But already there's no comparison.

In the 1980s, the energy produced by a new wind plant cost 57 cents a kilowatt-hour. Today, in windy locations, it's 2.1 cents (if you remove all subsidies, it's 4 cents). That's a 94 percent decrease in price. Over the next decade, experts predict that number will be sliced in half, bringing us "one cent wind" by 2030.

Solar is an even bigger story. Over the past forty years, there has been a three hundredfold reduction in the cost of making a solar panel. Forget kilowatt-hours for a moment. In 1977, generating one watt of power from a solar panel cost $77. Today, it's 30 cents, or a 250-fold reduction in price. "This price-performance curve is like nothing we've ever seen in energy," explains Ramez Naam, the head of Energy, Climate and Innovation for Singularity University. "The explosion of solar is almost like a digital transformation in the most fundamental category of infrastructure."

This upsurge helps explain why Peabody Coal, the largest private coal company in the world, recently filed for bankruptcy. It's par for the course. Over the past decade, coal stocks have dropped 75 to 90 percent as eight of America's largest coal companies have filed Chapter 11. Asia has also joined America in this anti-coal rampage. In 2016 alone, China canceled the construction of 160 plants. India did something similar the following year, killing $9 billion in ongoing projects in a single month.

As coal dies, renewables take their place. The largest coal plant in North America, Nanticoke Generating Station in Ontario, Canada, recently became a solar farm. The United Kingdom now generates more energy from zero-carbon sources than coal, which is saying something because it was coal that helped unite that kingdom in the first place. According to research conducted by the Carbon Disclosure Project, over a hundred major cities got 70 percent of their energy from renewables in 2017. The same year, Costa Rica spent 300 days running entirely off renewables, and other countries aren't that far behind. In total, 8 percent of the world's electricity now comes from solar and wind, and it costs less to build a new wind farm or solar plant than it does to operate an existing coal plant.

The upshot? Very cheap energy.

And cheap everywhere. In sunny parts of the US, solar costs 4.5 cents a kilowatt-hour. In India, where coal was supposed to dominate for most of this century, it's 3.8 cents. Abu Dhabi: 2.4 cents—which, when this contract was signed, was the lowest energy cost in history. Then Chile beat it with 2.1, and Brazil beat that with 1.75. In equatorial countries, essentially the places where the majority of people without electricity live, solar has become the cheapest form of energy available. More importantly, the poorest countries in the world are also the sunniest, which is going to completely invert the traditional power paradigm.

And there's more to come. Materials science is now merging with solar, changing the way we build panels and the way those panels perform. Take "quantum dots," essentially nanoscale chunks of semiconductor material that are starting to show up in solar cells. The big deal is energy conversion. A typical solar cell turns one photon of sunlight into a single electron of energy, meaning, today, in a very high-end panel, around 21 percent of incoming sunlight becomes outgoing energy. Quantum dots, meanwhile, triple this output, turning a solo photon into a trilogy of electrons, raising that conversion rate to 66 percent.

Technology is not only making solar more powerful, it's also making it more affordable. Right now, two-thirds of solar's price comes from soft costs—land, maintenance, sun tracking, effectively everything that's not the panel. Already, companies are using drones to monitor solar and wind farms and built-in sensors to track panel trouble before it starts. But we're not long from deploying robot technicians for solar and wind farm installation and maintenance, and using AI to supervise those technicians.

Finally, the reason we've been discussing solar and wind together is because these technologies too are converging—and with a huge upside. "The wind tends to blow when the sun doesn't shine, and vice-versa," says Naam. "That's true on an hour-by-hour basis, and on a season-by-season basis. Combining wind and solar on the same energy grid is a bit like adding 1 plus 1 and getting 3. If this existed in the US, we could, right now, meet 80 percent of our power needs."

But the most important point is also the most obvious one: Sunlight is free. And abundant. Every 88 minutes, 470 exajoules of solar energy hit our planet, which is as much as humanity consumes in a year. In 112 hours—or just less than five days—we get 36 zettajoules of energy, or what's contained in all proven oil, coal, and natural gas reserves on Earth. If we could capture just one-one-thousandth of that bounty, we'd have six times as much energy as we use today. And while the numbers are different, the same is true for wind. When it comes to energy, it's not about scarcity, it's about accessibility—which is the exact kind of problem that exponential technology has a history of solving.

The Story of Storage

If we're going to bring renewables to scale, we'll need to store energy. For emergencies, for peace of mind, for times when the wind won't blow and the sun won't shine, batteries are critical. But we're gonna need a lot of batteries.

Recently, California decided to source 100 percent of its electricity from renewables by 2045. To hit this goal, according to the Clean Air Task Force, the state needs 36.3 million megawatt-hours of energy storage. How much does it have today? About 150,000 megawatt-hours. In other words, California is .4 percent on their way.

Lithium-ion batteries have been everybody's initial solution to this problem. An exponential technology, these batteries have been dropping in price for three decades, plummeting 90 percent between 1990 and 2010, and 80 percent since. Concurrently, they've seen an elevenfold increase in capacity. But producing enough of them to meet demand has been an ongoing problem.

Enter the Gigafactory—Tesla's attempt to double global lithium-ion battery production. Located outside of Reno, the Gigafactory churns out twenty gigawatts of energy storage per year, marking the first time we've seen lithium-ion batteries produced at scale. A second Gigafactory has been built in Buffalo, a third in Shanghai, and a European location is under consideration. While it remains to be seen, Elon Musk has calculated that one hundred Gigafactories could manufacture enough storage for our planet's needs.

Tesla has also shown their batteries work at scale. In a 2018 project to upgrade a solar/wind farm in Australia, Tesla built the largest battery facility ever—one hundred megawatts of storage—in fewer than a hundred days. What's the big deal? First, we can now build fully integrated solar/wind/battery plants that produce energy at a price cheaper than coal. Second, we can do this over the course of a summer.

These developments caught the attention of other automobile companies. Renault started building home energy storage based on their Zoe batteries, BMW's 500 i3 battery packs are being integrated into the United Kingdom's national energy grid, and Toyota, Nissan, and Audi have all announced pilot projects. Despite this double-down, lithium-ion batteries are only part of our story.

Flow batteries are the next part. While lithium-ion batteries store energy in solids like metal, flow batteries store energy in liquids like molten salt. As lithium is a scarce resource found in dry climates—and mining it requires a half-a-million tons of water for every ton of lithium—replacing it with cheap and abundant salt is a useful substitution.

Flow batteries also serve different needs. Because lithium-ion batteries are lightweight and portable, they're perfect for mobile technology. The downside is durability. A typical lithium-ion battery can handle a thousand charge cycles. Flow batteries are the opposite. They're big and bulky, but can hold a charge for five thousand to ten thousand cycles, lasting decades without need for replacement. This makes them perfect for large utilities, data centers, and microgrids. San Diego, for example, as part of California's effort to scale up renewables, recently installed a flow battery that stores two megawatts of electricity, or enough to power a thousand homes for four hours.

Cost remains an issue. Flow batteries are currently more expensive than lithium-ion batteries, yet they're about to get a lot cheaper. Form Energy, with funding from Bill Gates's Breakthrough Energy Ventures, is working on an aqueous sulfur flow battery that costs one-fifth of its lithium-ion equivalent.

Dozens of different storage options are hitting the market as well. Companies such as Hydrostor are pumping compressed air into tanks and underground holding facilities, creating batteries that cost about half of traditional systems and last for over thirty years. There are also flywheel, thermal energy, and pumped hydroelectric storage systems coming online.

Materials science is further aiding the cause. Researchers at MIT are using carbon nanotubes to create "ultra-capacitors" that increase battery capacity by as much as 50 percent. And there's much more to come.

So the challenge isn't generating energy from renewables or storing the energy that gets generated, it's doing all this worldwide. This isn't just about building Ramez Naam's continent-wide smart grid, it's about building one of them on every continent. It's resource management on a global level, because, like it or not, when it comes to the environment, we really are in this together.

Electric Cars Are Gaining Speed

The final piece in the energy puzzle is transportation. In America, fueling our cars and trucks accounts for one-fifth of our total energy budget. Adding in planes, trains, and ships produces 30 percent of US greenhouse gas emissions. Globally, it's a slightly smaller 20 percent. While self-driving cars—which are predominantly electric—will lessen this burden, most experts feel this transition won't happen fast enough to keep global warming beneath two degrees.

To bring those numbers down, regulators everywhere have been pressuring the auto industry, announcing future bans on the sale of gas and diesel engines. Germany, the fourth largest carmaker in the world, was first down this road. In 2016, they announced the phaseout of the internal combustion engine by 2030. The following year, Norway sped past Germany, with their ban taking hold in 2025. Norwegians really got behind the idea, making 52 percent of their 2017 new car purchases electric. By comparison, it was 2.1 percent in the US in 2018.

India is also on board, aiming to be fossil fuel free by 2030. China, the biggest car market on the planet, is also considering a ban, with China's Volvo spearheading the effort, having already stopped production on everything but electric vehicles. Meanwhile, France, Germany, Denmark, Sweden, Japan, the Netherlands, Portugal, South Korea, Costa Rica, and Spain have all set official targets for electric car sales.

And where there's green there's green—which is to say, sensing a major market shift, pretty much every major car company now has an electric vehicle in production and/or for sale. The number of models available has grown from a paltry two in 2010 to a consumer-friendly forty-one in 2019. Ford alone is spending over $11 billion to electrify 40 vehicles by 2022. Daimler is outspending Ford, shelling out $11.7 billion on ten pure electric cars and forty hybrid models. But the largest single investor is Volkswagen, dropping $40 billion to electrify 40 vehicles by 2030. In total, global automakers have already plunked down more than $300 billion in investments.

A lot of this cash has been spent on batteries. Beyond teaming up with Tesla on the Gigafactory, Panasonic has also teamed up with Toyota to develop new battery technology, while Porsche and BMW are collaborating on ultra-fast charging stations. Volkswagen has invested in the startup QuantumScape, whose next generation solid state batteries are cheap, light, and—unlike their lithium-ion cousins (and much to the TSA's relief)—unable to catch fire. Plus, their energy density should provide a threefold improvement, putting the range of an electric vehicle very close to that of a gas-powered car.

Range, though, remains an issue. Today, most electric vehicles get about 200 miles a battery, but that number is trending upward. Range has been increasing by 15 percent a year for almost a decade. By 2022, your average mid-range model will get 275 miles on a full charge, while higher-end models will be closer to 350–400, which is the average range of a gasoline-powered car. By 2025, the year solid state batteries are supposed to hit the market, vehicles will push into the 500-mile range, or what many feel is required for widespread adoption.

The next piece in the electric puzzle: charging times. Your average gas station fill up takes less than ten minutes. The average EV fill up, depending on charger type, can stretch on for hours. But market forces and converging technology have sped things up considerably. That aforementioned Porsche and BMW collaboration, for example, has produced a four-hundred-megawatt charger that works twenty-five thousand times faster than the average smartphone charger. It pushes one hundred kilometers (sixty-two miles) of charge into a car battery in three minutes, and can take that same battery 10 percent charged to 80 percent charged in less than fifteen minutes.

The Tel-Aviv startup StoreDot has taken things even further. They've leveraged new materials to develop a lithium-ion "flash battery" that charges as fast as a super-capacitor, but discharges as slowly as a normal battery. A five-minute charge gives you three hundred miles of range. That's sixty miles a minute, or the rough equivalent of older model gas pumps.

Charging station availability is the final piece. Estimates vary, but most figure there are about 150,000 gas stations in America. Each averages eight pumps, for a national total of 1.2 million. By comparison, there are only sixty-eight thousand EV charging units in the US today. But these numbers are misleading.

They don't include residential charging, which is the number one charging location for EVs. Nor do they account for ChargePoint, a company that has raised over $500 million to build 2.5 million charging ports by 2025, half in Europe, half in the US. If successful, ChargePoint could put charge port availability on par with gas pump availability.

Which brings us to another of the World Economic Forum's top five dangers: extreme weather. In 2017, the average American home ran on 29.5 kilowatt-hours a day, while the average Tesla Model-S has an 85 kilowatt-hour battery pack. In a pinch, this means a fully charged Model-S could power three American homes for almost twenty-four hours. So if a hurricane takes out South Florida, a fleet of Teslas can be the emergency backup system. With an AI-driven smart grid, electric vehicles become nodes in a national network, a mobile fleet of backup generators to prepare for the extreme weather to come.

Biodiversity and Ecosystem Services

To round out our look at the more significant environmental threats we now face, we have to investigate species extinction and ecosystem collapse. The combination of climate change, deforestation, pollution, overfishing, and more, have created a giant biodiversity crisis. On a bad day, according to the UN, two hundred species go extinct. Forty percent of all insect species are in decline. Our closest relatives—chimps, apes, really all the members of the primate family—are now endangered. By century's end, at the current rate, 50 percent of all the large mammals will be gone.

The story might be worse in the ocean, where three-quarters of all the coral reefs are already in jeopardy. These reefs are home to 25 percent of the world's biodiversity, which both supports the livelihood of over 500 million people and produces 70 percent of the oxygen in the atmosphere. Yet, by 2050, if nothing changes, 90 percent of those reefs will be gone. And the picture isn't any brighter if you move out into the rest of the ocean. By 2100, a staggering 50 percent of all marine life will have disappeared.

Biodiversity is foundational to the health of our ecosystems and to the health of ecosystem services, which are all of the things that the planet does for us that we cannot do for ourselves. This includes oxygen production, food production, wood production, pollination services, flood protection, climate stabilization—thirty-six in total. And due to the loss of biodiversity, 60 percent of these services are critically degraded and unsustainable in the long term.

So how do we protect biodiversity and preserve ecosystem services? There is no simple solution, but we want to highlight five developments that are helping turn the tide.

Drone Reforestation: On land, forests are biodiversity hotspots, which is also why deforestation is one the largest drivers of extinction. The scale of destruction is vast. Every year, we lose 18.7 million acres of forest, or a swatch as big as Panama. Since trees are a major carbon sink, deforestation also accounts for 15 percent of total annual greenhouse gas emissions. So how do you combat industrial-scale deforestation? With industrial-scale reforestation.

Enter BioCarbon Engineering, a British company founded by ex-NASA employees that has developed AI-guided tree-planting drones. These drones first map an area to identify prime planting locations, then fire seed pods tucked inside of biodegradable missiles into the ground. The pods contain a custom-designed gelatinous growth medium that acts as a shock absorber to cushion impact, then a nutrient-dispenser to speed plant growth. A single pilot can fly six drones at once, planting a staggering one hundred thousand trees a day. A global army of ten thousand drones, which is what BioCarbon intends to build, could replant a billion trees a year.

Reef Restoration: Coral reefs are the forests of the ocean, so if we want to restore ocean health, we have to fix our reefs. There are around a half-dozen coral-regrowth technologies under development, but Dr. David Vaughan, a marine biologist with the Mote Tropical Research Laboratory, is pioneering some of the most exciting work. Borrowing tissue engineering techniques, Vaughan has figured out how to regrow one hundred years' worth of coral in under two years. And while normal coral will only spawn once it reaches maturity—something that can take twenty-five to a hundred years—Vaughan's corals reproduce at age two, giving us, for the first time, a way to radically replenish our reefs.

Aquaculture Reinvention: Fishing is one of the largest drivers of ocean wildlife decline, Right now, one-third of all global fisheries are stretched beyond their limits. Better fishery management is critical—but why manage when you can grow. The same tissue engineering techniques that allow us to produce steak from stem cells allows us to grow mahi-mahi, bluefin tuna, etc. In fact, there are now six different companies pursuing exactly this goal, with everything from cultured salmon to lab-grown shrimp heading for our menus.

Agricultural Reinvention: Plants and animals need room to roam, enormous stretches of pristine, uninterrupted habitat, both terrestrial and aquatic. Right now, 15 percent of the Earth is protected wildlands. To stave off what's now known as "the Sixth Great Extinction," Harvard's E. O. Wilson and other experts believe that half the planet might be required. Which raises a critical question: Where do we find that land?

In a nutshell, by coupling reforestation and restoration with the reinvention of agriculture. Roughly 37 percent of the globe's landmass and 75 percent of its freshwater resources are devoted to farming: 11 percent for crops, the rest for beef and dairy. However, these totals are shrinking. Not only are farmers abandoning their land in record numbers, but all of the innovations described in the Future of Food chapter—cultured beef, vertical farming, genetically engineered crops, etc.—allow us to harvest much more from far less. So, simple idea, let's give this extra land back to nature.

Closed-Loop Economies: Pollution is another top five threat we now face. A 2017 study conducted by the medical journal the *Lancet* estimates that pollution kills 9 million people a year and costs almost $5 trillion. The impact might be worse on nature. Obviously, greenhouse gas pollution is the biggest danger, but chemicals in our rivers, plastic in our oceans, and particulates in our air, are choking the life out of our planet.

So what can be done? Shifting from a petroleum-based economy to one powered by renewables will help, but more is needed. Arguably the biggest bat is zero-to-zero manufacturing. This process allows companies to completely remove waste rather than managing it via landfill. The list of companies now going this route is growing: Toyota, Google, Microsoft, Procter & Gamble, and more. Not only is this good for the environment, it's good for the bottom line. GM recently reported they've saved $1 billion over the past few years with their 152 zero waste facilities.

When we started this chapter, we highlighted the World Economic Forum's top five threats—water crises, climate change, biodiversity loss, extreme weather, and pollution. We've addressed all five individually, yet they're not individual problems.

Extreme weather results from climate change, but its effects are magnified by other issues. Consider Myanmar's Irrawaddy River Delta, once a biodiversity hotspot that was home to one of the largest mangrove forests on Earth. Over the past few decades, nearly 75 percent of that delta has been deforested, shutting down basic ecosystem services such as flood protection. When a cyclone hit the area in 2008, over 138,000 people died—with much of that devastation due to the loss of mangroves as a barrier.

But just as these are overlapping problems, there are overlapping solutions. Right now, Biocarbon Engineering's drones are replanting a portion of the Irrawaddy Delta that is double the size of New York's Central Park. Not only will this provide much needed habitat for wildlife, it will reboot ecosystem services such as flood protection. Plus, as mangrove forests store three times the amount of carbon as regular forests, this reforested delta becomes an invaluable tool in the fight against global warming.

In other words, the web of life is not a metaphor. Everything impacts everything impacts everything. The solutions we've highlighted all solve multiple problems at once. But we must be all in and right now. Stanford researchers give us three generations to halt species die-off before ecosystem services shut down in earnest. The Intergovernmental Panel on Climate Change estimates we have twelve years to halt global warming at 1.5 degrees. Yet we already have the technology required to meet these challenges, and thanks to convergences, it will only continue to improve. Our innovations may have caught up with our problems. Collaboration is the missing piece of the puzzle. If we're going to make the shift to sustainable at the speed required, then we the people are both the obstacle and the opportunity.

Economic Risks: The Threat of Technological Unemployment

When it comes to dangers heading our way, the environment gets top billing, but it's lately been sharing the limelight with automation. Robots and AI, our headlines increasingly declare, are coming for our jobs. In recent years, major consultancies such as McKinsey, Gartner, and Deloitte have all issued reports saying technological unemployment is unavoidable. One Oxford University study found 47 percent of all US jobs are threatened over the next few decades, and that number could be as high as 85 percent in the rest of the world.

Yet the facts tell a different story. Consider the job market, which is one of the first places to look for signs of this coming robopocalypse.

Except, as journalist and author James Surowiecki wrote in a 2017 article for *Wired*:

> Unemployment is below 5 percent, and employers in many states are complaining about labor shortages, not labor surpluses. And while millions of Americans dropped out of the labor force in the wake of the Great Recession, they're now coming back—and getting jobs. Even more strikingly, wages for ordinary workers have risen as the labor market has improved. Granted the wage increases are meager by historical standards, but they're rising faster than inflation and faster than productivity. That's something that wouldn't happen if human workers were on the fast track to obsolescence.

History tells a similar story. Theoretically, workers have been on the fast track to obsolescence since the Luddites took sledgehammers to industrial looms in the early 1800s. In 1790, 90 percent of all Americans made their living as farmers; today it's less than 2 percent. Did those jobs disappear? Not exactly. The agrarian economy morphed, first into the industrial economy, next into the service economy, now the information economy. Automation produces job substitution far more than job obliteration.

Even when there's automation, this doesn't always create the dire results we expect. Consider automatic teller machines (ATMs). When they were first rolled out in the late 1970s, there were serious concerns about bank teller layoffs. Between 1995 and 2010, the number of ATMs in America went from one hundred thousand to four hundred thousand, but mass teller unemployment wasn't the result. Because ATMs made it cheaper to operate banks, the number of banks grew by 40 percent. More banks meant more jobs for human bank tellers, which is why bank teller employment actually rose during this period.

The same thing is true for textiles, as journalist T. L. Andrews pointed out in *Quartz*: "Despite the fact that 98 percent of the functions of making materials have now been automated, the number of weaving jobs has increased since the nineteenth century." The same thing is true for paralegals and law clerks, two professions predicted to suffer job loss as a result of AI. Yet discovery software, introduced into law firms in the 1990s, has actually led to the inverse. Turns out, AI is so good at discovery that lawyers now need more humans to sift through the deluge—so paralegal employment has increased.

Productivity is the main reason companies want to automate workforces. Yet, time and again, the largest increases in productivity don't come from replacing humans with machines, but rather from augmenting machines with humans. "Certainly, many companies have used AI to automate processes," explained Accenture's James Wilson and Paul Daugherty in the *Harvard Business Review*, "but those that deploy it mainly to displace employees will see only short-term productivity gains. In our research involving 1,500 companies, we found that firms achieve the most significant performance improvements when humans and machines work together." BMW, for example, saw an 85 percent increase in productivity when they replaced their traditional—that is, automated—assembly line process with human/robot teams.

Also worth pointing out is that every time a technology goes exponential, we find an internet-sized opportunity tucked inside. Taking advantage of these opportunities requires adaptation—which demands workforce retraining—yet the end result is a net gain in jobs. Look at the internet itself. According to research done by McKinsey, in thirteen countries stretching from China and Russia to the US, the internet created 2.6 new jobs for every 1 it destroyed. Overall, in each of these thirteen countries, the Web's rise contributed 10 percent to GDP growth, and that number is still increasing.

Make no mistake, certain jobs are heading for extinction. While experts predict technological unemployment will have a greater impact in the 2030s, the next decade could see whole categories of work start to become memories. Robots are coming for everyone from truckers and taxi drivers to warehouse workers and retail employees. Amazon Go might not be the end of all cashiers, but in grocery stores, convenience stores, and gas stations, people will be absent more than present. The real question is will there be enough time to retrain our workforce before these effects go wide.

The answer appears to be yes. For example, Goldman Sachs recently made headlines with a study showing that autonomous vehicles will claim three hundred thousand driving jobs a year. What garnered less attention was their claim that we have twenty-five years to make this transition. Equally important, every educational advance—from VR-accelerated learning environments to AI-directed learning curricula—will make retraining easier, quicker, and more effective. Finally, as artificial intelligence becomes our user-friendly interface with technology, we're going to see a shift in the skills required for retraining. For a host of jobs, technological fluency and agility will replace deep skills mastery.

Once again, it comes down to whether we can collaborate or not. In July 2018, with 6.7 million unfilled jobs in the US, labor shortages reached a record high. Not only are the jobs there, they're there in record numbers. The ability to rapidly retrain our workforce to fill those jobs—that's the challenge we have yet to meet.

Existential Risks: Vision, Prevention, and Governance

In 2002, a relatively unknown Oxford philosopher named Nick Bostrom published a paper in the *Journal of Evolution and Technology*. In a few years, Bostrom would vault to geek-fame because of his "Simulation Hypothesis," which convincingly argues that we're living inside the Matrix. However, this earlier paper also caused a stir, mainly because it scared the crap out of nearly everyone who read it.

Bostrom's paper described a new kind of threat, which he dubbed an "existential risk," aka a "global catastrophic risk," but of a slightly different variety. Traditionally, "global catastrophic risks" have referred to everything from planet-killing asteroids to all-out nuclear war. But Bostrum wanted us to know that there was a new terror in town. Exponential technology, in his opinion, had a bad habit of becoming existential risk.

Nanotech run rampant—aka, Eric Drexler's "grey goo"—is one familiar example. Another is a pissed off AI waking up, hacking NORAD, and going DEFCON 666 on the entire world. There's also genetically modified organisms overrunning ecosystems, cyberterrorists playing good night New York with the power grid, or biohackers playing goodbye San Francisco with weaponized Ebola. These are the horrors that go bump in the expo-tech night. And this is Bostrum's dark point: We're in for a bumpy ride.

Yet, are we certain?

This is a contentious question. Sure, thought leaders like Elon Musk and the late great Steven Hawking have been exceptionally vocal about existential dangers, and institutions as august as Oxford and MIT have formed departments devoted to their study, but opinions remain all over the map. Trying to find accurate odds about our odds of survival is an exercise in futility. Despite this mess, a handful of consensus opinions have started to emerge. These are less solutions than categories of solutions: Vision, Prevention, and Governance.

Vision

Vision is about time horizons, how far we choose to look into the future. Our brains emerged in an era of immediacy, so we're a shortsighted species. How to avoid being eaten by a tiger—today. How to find enough food to feed my family—today. If there was any long-term thinking, it was of the *how do I find someplace warm to winter* variety. In other words, evolution shaped our time horizons to see about six months into the future.

Of course, we evolved ways to extend this perspective. Delayed gratification is the psychological term, and one distinguishing characteristic of our species is the ability to delay gratification beyond the limits of lifespan. Religions that shape behavior today by promising an afterlife tomorrow rely on this mechanism. No other animal can do this.

But we seem to be losing this talent. "Civilization is revving itself into a pathologically short attention span," writes Stewart Brand in an essay for the Long Now Foundation. "This trend might be coming from the acceleration of technology, the short-horizon perspective of market-driven economics, the next-election perspective of democracies, or the distractions of personal multi-tasking. All are on the increase. Some sort of balancing corrective to the short-sightedness is needed."

The corrective Brand came up with was his aforementioned Long Now Foundation, an organization most famous for constructing a clock that's hidden in a cave in the hinterlands of Nevada's Great Basin National Park. The clock's designed to keep time for ten thousand years, but its real purpose is psychological. It's built to make us think about ten-thousand-year time horizons. The organization's ultimate goal is to get people to understand that if you're trying to protect against existential risks—then you need to think long term.

Prevention

So how does thinking long term work in the real world? Prevention, our second category. One example is the Netherlands. Much of the nation sits below sea level, so it's Europe's most climate change–threatened locale. But rather than seeing rising tides as a problem in need of a quick fix—like larger seawalls that will, in turn, require short-term maintenance and eventual replacement—the Netherlands is being long-term proactive. "From a Dutch mindset," explained Michael Kimmelman in the *New York Times*, "climate change is not a hypothetical, or a drag on the economy, but an opportunity. . . . [The] Dutch are pioneering a singular way forward. It is, in essence, to let water in, where possible, not hope to subdue Mother Nature: to live with water, rather than struggle to defeat it. The Dutch devise lakes, garages, parks and plazas that are a boon to daily life but also double as enormous reservoirs for when the seas and rivers spill over."

Another example sits at the convergence of AI, networks, sensors, and satellites. Here, we gain the ability to develop global threat detection networks far more sophisticated than anything that exists today. Suggestions run the gamut, such as global food web monitoring to protect against catastrophic famines or terrorist attack; atmospheric sniffers that hunt for everything from plague-causing pathogens to the scent of nuclear materials; and rogue-AI detectors—essentially AI built to hunt for rogue AI.

And while all of this may seem outlandish, consider planet-killing asteroid detection. Two decades ago, this idea seemed somewhere between conspiracy theory and Hollywood thriller. Today, it's the "Sentry System," designed by NASA's Jet Propulsion Laboratory for "earth impact monitoring," and NASA's DART project, our first asteroid deflection mission designed for planetary defense.

Less futuristic but no less otherworldly, we've been using satellite imaging to track wildfires for a little while now. In 2018, NASA started training AI to interpret the data. After a year, their neural nets could detect forest fires from space with 98 percent accuracy.

Other researchers are working on what to do about the fires we detect. Firefighting drones are already in development. Before decade's end, it's not ridiculous to assume that space-based forest fire–spotting AIs will be communicating with autonomous fire-fighting drones down here on Earth—or an early step toward the dematerialization of emergency services.

This kind of thinking is mandatory. Even without technological advancement, the Earth is a living system where change is a constant. Originally, our atmosphere was a delightful combination of methane and sulfur, until a poison gas called oxygen came along and ruined everything. The dinosaurs enjoyed a spot as the uber-dominant creature on our planet until the dinosaurs enjoyed a spot in our museums, celebrating their onetime uber-dominance. In a turbulent world, unless we'd like to join the dinosaurs, we need to master this art of prevention.

Governance

In a rapidly changing world, prevention might be key to defeating existential risks, but adaptability and agility are the ultimate measures of prevention. Yet this is not exactly how society is organized. The majority of our organizations and institutions were built in another era, at a time when success was measured in size and stability. For most of the last century, standard metrics for business success were number of employees, ownership of assets, that sort of thing.

In our exponential world, agility beats stability, so why own when you can lease? And why lease when you can crowdsource? Airbnb built the largest hotel chain in the world, yet doesn't own a single room. Uber and Lyft have all but replaced cab companies in every major metropolis yet don't own a single taxi. And this level of flexibility, while now a requirement in business, is equally necessary in governance, which is our third and final category.

Modern ideas about government emerged about three hundred years ago, in a post-revolutionary world, when a desire for freedom from tyranny went hand in hand with a desire for stability. Thus, modern democracies are multi-house systems, a redundancy created to provide checks and balances. To fight tyranny and instability, these systems are designed to change slowly and democratically.

Our exponential world demands much faster reaction times.

Since 1997, the tiny Baltic state of Estonia has become a pioneer of e-governance, or the digitalization of what's traditionally been the most sluggish and recalcitrant sector on Earth. The goal is to massively speed up reaction times. Have a problem you need the government to solve? In almost any country in the world this means long lines, red tape, and big headaches. In Estonia, 99 percent of all public services are online, with user-friendly interfaces. Citizens pay taxes in fewer than five minutes, vote securely from anywhere in the world, and access all of their health information from a decentralized, blockchain-protected database. In total, the nation estimates they've reduced bureaucracy so much that they've saved *eight hundred years* of work time.

Encouraged by Estonia's example, governments around the world are going digital. And startups are trying to help. OpenGov turns the morass of government finance into a series of easy-to-read pie charts; Transitmix allows for real-time, data-driven transportation system planning; Appallicious created a disaster-assistance dashboard to coordinate emergency responses; Social Glass makes government procurement fast, compliant, and paperless.

The larger tech companies are also in on the action. Alphabet's Sidewalk Labs, for instance, is collaborating with the Canadian government on Quayside. In this smart community slated for Toronto's industrial waterfront, robots deliver the mail, AI uses sensor data to manage everything from air quality to traffic flow, and the entire cityscape is "climate positive," that is, built to green standards and sustainably powered. But what makes this project more than just interesting real estate news is that all of the software systems developed for Quayside will be open sourced, so anyone can use them, speeding up progress in smart cities everywhere.

Will any of this—from NASA's asteroid detection plans to the Netherlands' water-friendly redesign to Estonia's nimble e-governance—be enough to de-risk exponential risk? The answer is somewhere between "not close" and "not yet." But there are three reasons for optimism.

First, technological empowerment. Five hundred years ago, the only people capable of addressing these sorts of global, grand challenges were royalty. Thirty years ago, it was large corporations or big governments. Today, it's all of us. Exponential technology gives small teams the ability to tackle large problems. Second, opportunity. One of the central points we made in our last book, *BOLD*, was that the world's biggest problems are also the world's biggest business opportunities. This means that every one of the risks we face, whether environmental, economic, or existential, is the basis for entrepreneurship and innovation. Third, convergence. We tend to think linearly about the dangers we face, trying to apply the tools of yesterday to the problems of tomorrow. But we're going to experience a hundred years of technological progress over the next ten years. In fact, many of the most powerful technologies we'll have at our disposal—artificial intelligence, nanotechnology, biotechnology—are only starting to come online. So yes, the threats we face might seem dire, but the solutions we already possess will only continue to increase in power.

CHAPTER FOURTEEN

The Five Great Migrations

Our species is a migratory one. Over the past seventy thousand years, we wandered out of Africa and kept on wandering. We climbed mountains, forged forests, swam rivers, crossed continents, sailed oceans, and, eventually, managed to work our way into every corner of the Earth. It was an exodus-driven influx of innovation. While we left the old and sought the new, we brought our ideas, technologies, and cultures along for the ride. And this process is not just how the Harlem Shake got to Hong Kong, it's how we—all of us—got to now.

This has not been an easy journey. A great many of our mass migrations began with people fleeing from danger, disaster, and all the unspeakable horrors we now know as "history." Yet, despite originating in strife and tragedy, in the long run, migration has a positive impact on culture. In their book *Exceptional People: How Migration Shaped Our World and Will Define Our Future*, Oxford's Ian Goldin and Geoffrey Cameron explain it this way:

> The history of human communities and world development highlights the extent to which migration has been an engine of social progress. By viewing our collective past through the lens of migration, we can appreciate how the movement of people across cultural

> frontiers has brought about the globalized and integrated world we inhabit today. . . . As people moved they have encountered new environments and cultures that compel them to adapt and innovate novel ways of doing things. The development of belief systems and technologies, the spread of crops and production methods, have often arisen out of the experiences of, or encounters with, migrants.

Migration, as Golding and Cameron recount, isn't just people on the march, it's ideas on the move. It is, as it has always been, a major driver of progress. Migration is an innovation accelerant.

A few years back, a Stanford economist named Petra Moser (now at NYU) decided to try to quantify the impact of this acceleration. It was something of a personal inquiry. "More than half of my colleagues at Stanford are immigrants," she once told reporters. "I [wanted] to find out how policies that alter the flow of such highly skilled immigrants affect science and innovation."

To answer this question, Moser and her team turned to an old rumor—that German Jews who fled Nazi Germany had an outsized impact on innovation in the United States. If true, it was an outsized impact produced by an outsized exodus.

The outpouring began in April of 1933, when Adolph Hitler passed the "Law for the Restoration of the Professional Civil Service," banning all "non-Aryans" from government employment. Tens of thousands lost their jobs: firefighters, cops, teachers, and, most important for this discussion, academics. Only two months after Hitler became chancellor, the writing was on the wall. Over the next decade, more than 133,000 German Jews fled for America. In context, it's as if every living soul in Charleston, South Carolina, relocated to Texas or, it would be, if the population of South Carolina also included Albert Einstein and five other Nobel Laureates.

To measure the impact of this influx, Petra Moser started with chemistry patents. Then, she expanded into nearly every technical field, measuring the number of patents applied for and received from 1920 to 1970, tracking the impact of migration through the records of more than a half-a-million inventions.

What did she find? That migration is an innovation accelerant on par with nearly every force we've so far discussed. In every field entered by German Jews, she found a 31 percent increase in patents. Back then, as anti-Semitism was rampant in the United States, a great many of these immigrants were prohibited from working in their chosen professions. When Moser and her team adjusted the data to account for this fact, they found that émigrés actually accounted for a stunning 70 percent patent increase.

While Moser's work confirmed the rumors, and gave us a different way to look at both the power of immigration and this extraordinary period in history, it's also worth noting what's not extraordinary—that migration drives innovation. This same pattern continues today. A 2012 study by the Partnership for a New American Economy, for example, found that three out of four patents issued to America's top ten patent-producing universities have at least one foreign-born inventor.

A different look at this same trend comes via "product reallocation," which describes the rate at which new goods and services enter the market and force old ones to exit, or what economist Joseph Schumpeter called "creative destruction." Far more than patents, researchers consider product reallocation the gold standard for innovative impact.

A few years ago, researchers from the University of California, San Diego, found a direct link between migration and this gold standard. By tracking the product reallocation rate for every American company that hired a highly skilled foreign-born worker between 2001 and 2014, they found a very clear signal. Companies with high-skilled foreigners saw an increase in both their rates of innovation and the impact those innovations had on the market: A 10 percent rise in skilled foreigners on the payroll led to a 2 percent increase in product reallocation. And this was true regardless of how much the company spent on research and development.

What's true for migration's impact on invention is also true for entrepreneurship. While much has been made about immigrants taking work away from citizens, the data shows the opposite. Rather than stealing jobs, they are far more likely to create new ones.

In America, immigrants are twice as likely to start a new business than natives, and are responsible for 25 percent of all new jobs. Between 2006 and 2012, 33 percent of venture-backed companies that went public had at least one immigrant founder. Among Fortune 500 companies, 40 percent were founded by immigrants or their children. In 2016, half of all unicorns—those rare startups valued at more than $1 billion—were founded by immigrants, and each provided at least 760 new jobs.

And why does this matter so much?

Two reasons. First, the challenges outlined in the previous chapter are going to require significant innovation. We'll need new ideas to counter environmental and existential risks and new jobs to replace the ones that robots and AI are about to make obsolete. To implement those ideas, we'll also need greater global collaboration and cooperation, and a deep empathy that crosses borders, cultures, and continents. And thanks to five of the greatest migrations the world has yet seen, we'll soon see all of this and more.

In this chapter, as we widen our view from the next decade to the century that follows, we're about to witness mass migration on a massive scale. In some cases, we're moving for familiar reasons—to avoid environmental disaster and chase economic opportunity—but in shorter time frames and greater numbers than anything yet seen. In others, we're crossing borders we've never crossed before. Moving off world and into outer space; moving out of regular reality and into virtual reality; moving, if the cutting-edge of brain-computer-interface development continues apace, out of individual consciousness and into collective consciousness, a technologically enabled hive mind, or, for those who speak "Trekkie," a kinder, gentler Borg.

So ladies and gentleman, fasten your seatbelts and please keep arms and legs inside the ride at all times. Migration is a serious accelerant. And over the next hundred years, thanks to five great migrations, we are about to play now-you-see-it, now-you don't with the world as we know it.

Climate Migrations

While the last chapter examined technological ways to mitigate climate change, this one acknowledges that our ability to implement these solutions at scale is nowhere near where it needs to be. And make no mistake, when the weather shifts, people shift with it.

Estimates of this impact are startling. And climbing. In 1990, the very first Intergovernmental Panel on Climate Change report warned that even a slight rise in sea levels could produce "tens of millions of environmental refugees." In 1993, Oxford scientist Norman Myers controversially updated the IPCC's prediction, arguing that climate change could displace as many as 200 million people by 2050. By decade's end, as Mark Levine explained in *Outside* magazine: "The weather [had] come to assume the shape of our collective anxieties, our fantasies about technology, nature, retribution, inevitability. . . . We have overstepped, we whisper, we have changed the weather. Now the weather is going to change us."

How much will it change us? In a 2015 meta-analysis of all available data, Climate Central, an independent group of leading scientists and journalists, reported that even if we manage to halt warming at two degrees, extreme weather will still displace 130 million people. If we don't? Climate Central's prognosis isn't good: "Carbon emissions causing 4 degrees C of warming—what business-as-usual points toward today—could lock in enough sea level rise to submerge land currently home to 470 to 760 million people."

To figure out what this level of displacement actually looks like, Climate Central also made a series of maps depicting the effects of global warming on every coastal nation and mega-city on the planet. Unless you're a fish, the news is not good.

With four degrees warming, in many of the world's mega-cities—London, Hong Kong, Rio, Mumbai, Shanghai, Jakarta, Calcutta, etc.—swimming becomes the fastest way to get from point A to point B. Entire island nations vanish forever. In America, twenty million people end up underwater. In Washington, DC, sea levels reach the Pentagon. And if you thought real estate in New York was expensive today, just wait until everything south of Wall Street disappears.

Beyond the deluge, global warming also puts the ancient nemesis of drought on the horizon. Drought drove us out of Africa some seventy thousand years ago, and is still driving us today. Syria has the highest number of refugees in the world, and it's partially because of drought. In Europe, even if we halt warming at two degrees, the Mediterranean will continue to dry, with Italy, Spain, and Greece being especially hard hit. "In other words," as journalist Ellie Mae O'Hagan wrote in the *Guardian*, "the Mediterranean countries currently trying to cope with migrants from other parts of the world may eventually have a migrant crisis of their own. One day there could conceivably be Italians and Greeks in Calais, as their own countries become even hotter and more arid."

In historical terms, the 1947 partitioning of India and Pakistan is considered the greatest forced migration in history, upending some 18 million people. Even if we put climate migration at the low end of the prediction spectrum—meaning two degrees warming and 130 million displaced—we're still looking at a global reshuffling seven times larger than anything seen before.

Yet climate migration is a peculiar kind of forced migration, as we ourselves are doing the forcing. The cost in both hard dollars and human suffering is much higher than we should be willing to pay. With a population of 38 million, Tokyo is the largest mega-city on Earth. Consider what it would cost to relocate fifteen Tokyos. Now consider that this is an entirely voluntary expense.

As our prior chapter explored, we have a great many of the strategies and technologies required to address climate change. Whatever it might cost us to implement these solutions, it's going to be a whole lot cheaper than finding new homes for 700 million people. Either way, in the long run, as the weather sends us hither and yon, the rate of innovation will, as it has always done, continue to climb.

Urban Relocations

The immense scale of climate migration—those 700 million on the move—represents the largest demographic reshuffle in history. Yet, measured against our second torrent, it's barely a trickle. As over the next few decades, damn near everyone is moving downtown.

Three hundred years ago, 2 percent of the world's population lived in cities. Two hundred years ago, it was 10 percent. But the Industrial Revolution's steam-powered punch forever altered those numbers. Between 1870 and 1920, 11 million Americans left the country for the city. In Europe, 25 million more crossed an ocean to settle, predominantly, in US cities. By 1900, 40 percent of the United States had urbanized. By 1950, it was 50 percent. By the turn of the millennium, 80 percent.

The rest of the world wasn't far behind. Over the past fifty years, in low- to medium-income countries, urbanization has doubled, sometimes tripled—think Nigeria and Kenya. By 2007, the globe had crossed a radical threshold: Half of us now lived in cities. Along the way, we got cities on steroids. In 1950, only New York and Tokyo housed 10 million residents, which is the figure required for "mega-city" status. By 2000, there were over eighteen mega-cities. Today, it's thirty-three. Tomorrow?

Tomorrow is when the numbers go crazy. In fact, we have a new word for the crazy, a "hyper-city," a locale with a population above 20 million. By comparison, during the French Revolution, the world's entire urban population was less than 20 million. By 2025, Asia alone will house ten, maybe eleven hyper-cities.

And we're going to need 'em.

By 2050, some 66 to 75 percent of the world will have urbanized. With over 9 billion expected by then, this is the growth spurt to end all growth spurts. It's an exodus three times larger than the one about to be produced by climate change, the real largest migration in history, a mass movement some 2.5 billion strong.

And when the masses move, they move masses.

By 2050, Tokyo loses its title, as Delhi is expected to become the world's most populous town. And China out-urbanizes India, adding three hundred new million-plus cities and two mega-cities. Africa just explodes. From Cairo through the Congo, the continent's urban population grows 90 percent by 2050. By century's end, Lagos, Nigeria, could be home to 100 million.

Add it all together, every week from now until 2050, a million people move downtown. University of Toronto urban studies professor Richard Florida calls this the "central crisis of our time." Like any crisis, this one brings both opportunity and danger.

First the upside.

From an economic perspective, cities are good for business. In 2016, the Brookings Institute examined the 123 largest metro economies in the world. While housing only 13 percent of the planet's population, they produced almost one-third of its economic output. The following year, the National Bureau of Economic Research took a second look at this relationship between productivity and population density. They found the same pattern: more people, more productivity.

London and Paris, for example, are significantly more productive than the rest of Britain and France. In America, our hundred largest cities are 20 percent more productive than all others. In Uganda, urban workers are 60 percent more productive than rural ones. Shenzhen's GDP, meanwhile, is three times larger than the rest of China.

Density also drives innovation. Santa Fe Institute physicist Geoffrey West discovered that every time the population of a city doubles, its rate of innovation, as measured in number of patents, increases by 15 percent. In fact, in West's research, no matter the city studied, as population density increases, so do wages, GDP, and quality-of-life factors like the number of theaters and restaurants.

And, as cities grow, they require less, not more, resources. Double the size of a metropolis, and everything from the number of gas stations to the amount of heat needed in the winter—only increases by 85 percent. Turns out, larger, denser cities are more sustainable than smaller cities, small towns, and suburbs. Why? Travel distances drop, shared transportation rises, and less infrastructure—hospitals, schools, garbage collection—is required. The result is that cities are cleaner, more energy efficient, and emit less carbon dioxide.

And smart cities could take this further. A 2018 McKinsey study found smart city solutions could reduce urban greenhouse gases some 15 percent, solid waste by 30 to 130 kilograms per person per year, and save water—some twenty-five to eighty gallons per person per day. In fact, using today's technology, we could achieve 70 percent of the UN's Sustainable Development Goals merely by transitioning to smart cities.

Now the downside: Calamity is a definite possibility. Unplanned urbanization is a fantastic recipe for crime, disease, the cycle of poverty, and environmental devastation. Yet, as this book makes clear, our tools are equal to those challenges. The tricky part is matching visionary technology with plain old vision—good governance and civic cooperation. Get it right and urbanization becomes one of the most effective tactics in our fight against many of today's pressing problems. Get it wrong? Then the largest migration in history will produce the biggest messopolis in history.

Virtual Worlds

By the numbers, the 12 million Africans uprooted by the slave trade, the 18 million people rerouted by the division of India and Pakistan, and the 20 million rearranged on Europe's chessboard in the years following World War II were history's three biggest forced relocations. Each was propelled by a familiar driver: economics (and depersonalization), religion, and politics, respectively. Each reshuffled the world. Yet their combined impact will soon be dwarfed by a new exodus, the first to be triggered solely by technology.

Our next migration begins with the flick of a switch.

Sometime over the next few years, someone, somewhere, will jack themselves into the Matrix and never look back. Welcome to the strangest mass exit yet encountered: our migration from normal reality into virtual reality.

Our bags are already packed. Globally, video games consume three billion hours a week. In America, digital media devours eleven hours a day. Internet gaming disorder is a recognized mental health condition, and tales of excess are myriad. In 2005, the BBC reported that a South Korean man died after spending fifty consecutive hours playing an online video game. His demise was the first of many. In 2014, the *Guardian* broke news of a couple who left their three-month-old baby to starve to death while they raised a virtual baby online at a local internet café. In Japan, there's even a word for it: *hikikomori*, the lost generation, the invisible youth, the nearly 1 million teenagers who have locked themselves in their rooms and only venture out online.

These people are migratory pioneers. They're setting up beachheads for virtual exploration. But over the next few decades, two factors will amplify this influx. Let's call them psychology and opportunity.

We'll start with psychology. While all previous migrations have been triggered by external drivers, or things happening in the world, this next one will be triggered by internal drivers, psychological drivers, or things happening in our brains. This next migration begins with our own addictive neurochemistry, against which there is no known defense.

Video games are addictive. At the root of this addiction is the thrill ride known as dopamine, one of the brain's primary pleasure drugs. We feel dopamine as engagement, excitement, a desire to investigate and make meaning out of the world. It's released whenever we take a risk, expect a reward, or encounter novelty. Once hardwired into a reward loop—meaning, once our brain establishes a link between an activity and dopamine—the desire to get more of this chemical becomes our overarching preoccupation. Cocaine, by way of comparison, is one of the most addictive substances on Earth, yet much of what it does is flood the brain with dopamine.

Video games are chock-full of risk, reward, and novelty—they're dopamine dispensers dressed up like joysticks. But it's not just video games. When your phone buzzes with a message, that urge to see what it says, that's dopamine too. The little rush of pleasure you get from checking that message, dopamine as well. Nearly all the major uses of the internet—gaming, surfing, social media, texting, sexting, and porn—are dopamine drivers. Yet none of these drive dopamine like VR.

Research shows that the immersive nature of the virtual environment spikes dopamine to heights typically unattainable by traditional video games or any other kind of digital media. While numbers vary slightly, most researchers believe that video games are truly addictive to about 10 percent of the population. Virtual reality will significantly increase that percentage. "Facebook is an addictive technological drug that, like every drug, gives people temporary pleasure and, ultimately, causes people to become psychiatrically ill," psychiatrist Keith Ablow recently explained in an article for *Fox News*. "Oculus Rift will make matters worse."

Yet dopamine is merely one of the brain's major reward chemicals. There's also norepinephrine, endorphins, serotonin, anandamide, and oxytocin to consider. All are massively pleasurable. Digital media isn't incredibly effective at producing any beyond dopamine, but the immersive nature of VR makes it able to trigger all six. It's the full cocktail of feel-good neurochemistry, hard drugs delivered by headset—and only the start of this story.

The next part emerges out of research into flow states. For those unfamiliar, flow is technically defined as "an optimal state of consciousness where we feel our best and perform our best." It's a state of peak performance, and part of what produces this state are all six of the brain's pleasure chemicals. This is why researchers consider flow to be one of the most addictive experiences available. Yet it's also one of the most meaningful. In over fifty years of research, the people who score off the charts for deep meaning and overall life satisfaction are the people with the most flow in their lives.

While video games can drive users into flow, the immersive nature of VR makes the technology significantly better suited for it. This means, as flow science and virtual reality continue to converge, we'll soon gain the ability to create an alternative reality that is both more pleasurable and more meaningful than regular reality. Now hold on to this idea for a moment, as we explore the opportunity side of this story. Three opportunities in particular: jobs, education, and sex.

On the jobs front, we already know there's economic possibility tucked inside of virtual reality. Second Life was the first virtual world. Back in 2006, *BusinessWeek* put on the cover real estate tycoon Anshe Chung, who, through deal-making *within* Second Life, became the first real-world millionaire to earn her fortune entirely in the virtual. We've seen similar profiteering inside of video games and social media, and virtual reality will bring more of the same. Which is to say, if robots and AI start to take a lot of our jobs over the next few decades, then the one-two punch of a shrinking job market in regular reality and an exploding job market in virtual reality makes for a potent migratory impetus.

The second trend is education. VR lets us create distributed, customized, accelerated learning environments. Whether it's our burgeoning global population looking for an education or our suddenly techno-unemployed population looking for retraining, we're seeing a force in the making. VR's ability to drive people into flow makes this even more potent, as the state amplifies our ability to take in and retain new information. Research conducted by the Department of Defense, for example, found soldiers in flow could learn 230 percent faster than normal. This is also why, in Ernest Cline's bestselling novel *Ready Player One*—where much of the world has already moved into VR—education was the primary driver of that migration.

Our final opportunity is sex. From VCRs to the internet, nearly every major communication technology has been driven forward by pornography. VR is clearly the next wave. Yet VR, augmented by haptics, makes pornography a multi-sensory experience. For the first time ever, you can look and you can touch—which brings with it a much bigger cocktail of addictive neurochemistry.

Plus, it's more than porn. It's also social media. Imagine VR Tinder, or the ability to sext your partner actual sensations. Stanford emeritus professor of psychiatry Al Cooper, who conducted one of the largest and most detailed studies of cybersex, described the Web as "the crack cocaine of sexual compulsivity." According to his research, two hundred thousand Americans are already digital sex addicts. Globally, that number creeps into the millions. If you consider that VR sex is better at producing dopamine than digital sex, then you begin to understand that there's another migratory driver here—one that mega-taps a primal evolutionary impulse.

Added together, our three largest migrations—the slave trade, the bifurcation of India and Pakistan, and the diaspora of post–WWII Europe—produced a combined 44.5 million exiles. Yet 321 million Americans already spend eleven hours a day online, and VR's neurochemical cocktail will definitely increase that figure. Now toss in serious human motivators like meaning, mastery, money, and sex, and the pull becomes much stronger. It adds up into another great migration, an exodus of consciousness, and one only now just beginning to get under way.

Space Migration

"[The] Earth is the cradle of humanity, but one cannot remain in the cradle forever," said Konstantin Tsiolkovsky in the late 1800s. Tsiolkovsky was a true visionary. Considered the father of space flight, he's the Russian scientist credited with first dreaming up airlocks, steering thrusters, multistage boosters, space stations, the closed-cycle biological systems needed to provide food and oxygen for space colonies, and much more. Over the course of his career, he published over ninety papers on these topics, envisioning nearly every aspect of what it would take to conquer this final frontier except, of course, what it actually took to conquer this final frontier: competition.

In the 1960s, what first drove us off-world was the messy showdown between ideologies and ideologues known as "United States v. Soviet Union." And it's competition that's still driving us today. Only now, while a handful of governments remain in this game—the US v. China, for example—the real story is a rivalry of tech titans: Jeff Bezos v. Elon Musk.

Each of these men has a deep desire to move us out of our cradle and into the stars, to open the space frontier and "back up the biosphere," creating a second human civilization in space in case things don't go so well here on Earth. And it's these dreams and this competition that has become a force of its own, both a great big push off-world and the only migration in history that comes with its own Twitter battle.

> @JeffBezos, Nov. 24, 2015: The rarest of beasts—a used rocket. Controlled landing not easy, but done right, can look easy. Check out video: bit.ly/OpyW5N

> @elonmusk, Nov. 24, 2015: Not quite "rarest." SpaceX Grasshopper did 6 suborbital flights 3 years ago & is still around.

We'll start with Bezos, whose passion for space began in high school. Both a child of the Apollo era and a serious *Star Trek* fan, Bezos's valedictorian speech focused on "a future where millions of people are living and working [off-world]," and closed with the line "Space, the final frontier, meet me there." In college at Princeton, Bezos was the chapter president of Students for the Exploration and Development of Space (SEDS). His time there overlapped with the late physicist Gerard K. O'Neill, the founder of the Space Studies Institute. In the early 1980s, O'Neill asked a key question of his students: "Is a planetary surface the best place for humans to live while they expand into the solar system?" Eventually determining that the answer was no, O'Neill instead proposed we build massive rotating cylinders, now known as "O'Neill colonies," manufactured from resources already outside the deep gravity of planets like Earth or Mars, manufactured, specifically, with materials gathered from the surface of the Moon.

These space lessons never left Bezos. After college, they helped carry him from Wall Street to Amazon as the first step in what he's jokingly called "a simple two-step plan. First make billions, then open the space frontier."

Once he did make billions, Bezos plunged them back into space. In 2000, he founded Blue Origin, committing a billion dollars a year to the project. His initial goal, then announced, was the construction of rockets capable of shooting people and payloads off Earth, into space, and, eventually, to the Moon—which he still believes is the best launch spot for our colonization of the cosmos.

"We were given a gift," said Bezos at a 2019 event in Washington, DC, "this nearby body called the Moon. It is a good place to begin manufacturing in space due to its lower gravity. . . . Getting resources from the Moon takes 24 times less energy to get it off the surface compared to the Earth. That is a huge lever."

As the next step, Bezos announced the Blue Moon Lunar Lander, which would travel to the Moon aboard his reusable New Glenn rocket, depositing 3.6 metric tons of rovers, cargo, and humans on the lunar surface. He also argued that we have no choice in this matter. "There is no Plan B. We have to save this planet. [Yet] we shouldn't give up a future for our grandchildren's grandchildren of dynamism and growth. [In space], we can have both."

Bezos then resurfaced O'Neill's work, proclaiming that Blue Origin's post–lunar landing vision was the development of O'Neill colonies, each supporting an independent population of 1 million, or one of the big drivers of our next great migration. "The Earth is the gem of our solar system," he explained. "It should be zoned residential and light industry. Heavy industry should be moved into space . . . where there's unimaginable room. . . . The solar system can support a trillion humans, and then we'd have a thousand Mozarts and a thousand Einsteins. Think how incredible and dynamic that civilization will be."

And even though there's a heathy rivalry here, Elon Musk doesn't disagree: "History is going to bifurcate along two directions: One path is we stay on Earth forever, and then there will be some eventual extinction event . . . the alternative is to become a spacefaring civilization and a multiplanetary species. I think the future is vastly more exciting and interesting if we're a spacefaring civilization and a multiplanet species, than if we're not."

Born in Pretoria, South Africa, Musk sold his first computer company at age twelve. After earning a degree from Wharton, then dropping out of Stanford's PhD program, he repeated his software success with first a $307 million sale of Zip2, next a $1.5 billion sale of PayPal. Finally, with what he considered sufficient resources to make a difference, Musk set out to pursue what he considered the two missions most critical to our survival: breaking our fossil fuel addiction with a thriving solar economy—i.e., his work with Tesla and Solar Cities—and making humanity a multiplanetary species. But unlike Bezos's lunar launch point for this migration, Musk's obsession has always been Mars.

In 2001, a year before the sale of PayPal, Musk came up with the idea of sending a plant—plant seeds, really—to Mars. In his "Mars Oasis" project, his spaceship would include a sealed chamber with an Earth-like atmosphere, a healthy collection of seeds, and a nutrient gel to speed their growth. "When you'd land," explained Musk, "you hydrate the gel and you have a little greenhouse on Mars."

Musk wanted to take photos of the plant growing on the surface of the red planet, an image so impactful he felt sure it would inspire the US government to fund missions to Mars and establish a permanent human colony there. But when he investigated the cost of buying the rockets required to send his greenhouse up, up, and away, he realized the available launch options were way too primitive and expensive to ever facilitate a human colonization of the stars.

To solve these problems, Musk founded SpaceX in 2002. In June 2008, after several spectacular failures and a close brush with bankruptcy, Falcon 1 got off the ground and into orbit. This success was followed by dozens more, each cheaper than the last. Next came reusability, a rocket that could take off and land without destroying itself and a longtime dream of the airspace industry. Finally came Falcon Heavy, the largest rocket on the planet, which, in early 2018, launched Musk's cherry-red Tesla roadster past Mars and on a trajectory toward the asteroid belt. Finally, SpaceX announced they would soon stop production of the Falcon vehicles—which lacked the oomph to send humans to Mars—and instead began work on "Starship."

Musk views the establishment of a Mars colony as a contingency plan for humanity and a problem to be solved this decade. Starship test flights are already under way and his stated goal is to have humans on the planet's surface before 2030, with a full city up and running by 2050. To achieve this, SpaceX has scheduled ten major launches between 2027 and 2050, one every twenty-two to twenty-four months, when the distance between the Earth and Mars is at its shortest.

The current plan goes like this: A Starship gets launched into orbit around Earth, then several tanker Starships would launch as well, meeting up with the first one to top up its fuel. From there, these rockets would head straight toward Mars, ferrying a crew and roughly a hundred passengers at a time. The cost per person for a one-way ticket? Musk thinks about $500,000, or, as he said: "low enough that most people in advanced economies could sell their home on Earth and move to Mars."

One thing for sure, whether it's Musk or Bezos who wins this particular space race, as Konstantin Tsiolkovsky also pointed out, a great many of the things that humanity has come to prize here on Earth—metals, minerals, energy, fresh water, prime real estate, endless adventure, lust, love, meaning, and purpose—they're all in near infinite quantities in space. And it is the quest to claim this treasure—today being played out by battling billionaires—that is blasting us out of our cradle and into the stars, on the front end of another of this century's great migrations, our first real foray across this, the final frontier.

Meta-Intelligence: Into the Borg

In 2015, Harvard chemist Charles Lieber was trying to solve a difficult problem in the new field of neuro-modulation. Over the past few decades, deep brain stimulators had been developed to help sufferers of Parkinson's disease. While the patient remains awake, a hole is drilled through the skull and a device that sends electrical pulses to areas of the brain responsible for movement is inserted. It's become almost a routine procedure. Over a hundred thousand devices have been implanted, and for patients who have exhausted all other medical options, deep brain stimulation remains the only way to improve motor control and lessen tremors.

Unfortunately, there are side effects. Weird side effects. The tendency to develop a compulsive gambling problem is the most frequent issue. Workaholics becoming couch potatoes overnight is another. Chronic depression is a third. The reason? Size.

If they had their way, neurosurgeons would like to impact the brain at a single neuron level, but today's deep brain stimulators are too big for this precision. Trying to target individual neurons with today's implants is, as MIT professor of materials science and engineering Polina Anikeeva pointed out in her 2015 TED Talk, "akin to trying to play Tchaikovsky's first piano concerto with fingers the size of a pickup truck."

Making matters more complicated, these devices require surgery to install and, as the brain treats them like foreign invaders, serious medicine afterward. There's also the issue of design. The body is a flexible 3-D environment, but most of today's brain implants—be they deep brain stimulators or otherwise—are inflexible 2D devices that have more in common with traditional silicon chips than anything that exists naturally in the body. In the squishy, hot, and wet of the brain, it's no wonder signals get crossed and side effects occur.

Charles Lieber, though, took a very different approach. To help regenerate bone, doctors often implant a "bioscaffold" into damaged areas to provide a support structure for new tissue to grow around. About five years ago, Lieber decided to try to build a microscopic bioscaffold made from electronics. He used photolithography to etch a four-layered probe one layer at a time, creating a nanoscale metal mesh with sensors capable of recording brain activity.

After rolling that mesh into a tight cylinder, Lieber sucked it up into a syringe, then injected it into the hippocampus of a mouse. Within an hour, the mesh had unfurled into its original shape, doing no damage to tissue along the way. The result: Mouse-Brain TV. Lieber could monitor the activity of the mouse's brain, in real time, in a living animal. The mouse's immune system treated the implant as friend not foe. Instead of attacking the mesh as a foreign invader, neurons attached themselves to it and began multiplying.

In a separate experiment, Lieber injected the mesh into the retina of a mouse, where it once again unfurled, doing no damage to the eye along the way. The result is a device that neither impairs sight nor blocks light, yet remains capable of recording mouse vision, at a single neuron level, across sixteen channels at once, for years on end. The work brought accolades to the Lieber Group and helped the technique spread like wildfire. How-to tutorials are available online. As are plenty of videos of Elon Musk describing the next step in the evolution of the idea, what he's termed "a neural lace," an injectable brain-computer interface, an "ultra-high bandwidth brain-machine interface to connect humans to computers."

Brain-computer interfaces (or BCIs) are the ultimate tale of convergence. They sit at the intersection of nearly everything in this book, including biotechnology, nanotechnology, and materials science—which, as we've seen, are rapidly becoming all the same industry. There's also quantum computing, which gives us the ability to model complex environments like the human brain, and artificial intelligence, which allows us to interpret what we've modeled. And high-bandwidth networks that allow us to upload neurological signals into the cloud. In fact, packed into this single advancement, we find most of our advancements.

If we consider our development of exponential technologies to be among the leading examples of human intelligence, then BCIs are the crowning achievement of those examples. They might also be a way to survive our own success, as, in the minds of many, BCIs are the upgrade we desperately need to fully participate in an AI-dominated world.

The leading proponents of this view are Elon Musk and Bryan

Johnson, who have both created companies, Neuralink and Kernel respectively, to speed development along. But everyone from Facebook to DARPA has gotten involved. Facebook wants neurotech that allows you to think instead of type, replacing the keyboard with the mind as the ultimate social media interface. DARPA sees BCIs as a next-generation battlefield technology, and wants one that can record 1 million neurons simultaneously while stimulating a hundred thousand. There are also a multitude of startups crowding into the space, from health and wellness to education and entertainment.

And progress has been made.

Over the past decade, utilizing EEG-based brain-computer interfaces—which require no surgery to install and instead just sit atop the head like a crown of electrodes—researchers have performed real miracles. BCIs have allowed paraplegics to walk again. Stroke victims, paralyzed for years, are regaining the use of their limbs. Epileptics have been cured of seizures. Quadriplegics can now control cursors with their mind. And, piling atop Dracula, flying cars, and personal robots in our list of childhood fantasies turned into realities, telepathy is also now possible.

Back in 2014, a team of Harvard researchers sent words from mind to mind via the internet. Technically known as "brain-to-brain communication," this was an example of the long-distance version—with one subject in France, the other in India. The researchers used a wireless, internet-connected EEG headset as their transceiver and a transcranial magnetic stimulator—which sends weak magnetic pulses into the brain—as a receiver. The subjects didn't exactly get thoughts, but rather could accurately read flashes of light that corresponded to the message.

And that was then. By 2016, we were using EEG headsets to play video games telepathically, and by 2018, we were piloting drones with our thoughts. The next step is figuring out how to seamlessly link our brains to the internet via the cloud—which is why Lieber's injectable mesh matters so much. The general thinking is that neurotech sitting atop the head won't be able to capture signals at any useful resolution, while devices that require surgery to implant—however minor the procedure becomes—remain too risky for widespread adoption. But an injectable neural-lace, to borrow Musk's term, solves these problems and then some.

And this brings us to our final migration, a sojourn out of our normal brain-based singular consciousness and into a cloud-based collective consciousness, both a hive mind and a reminder that the greatest journeys are often inward toward our psyche, rather than outward toward the stars. Economic considerations alone, as both Elon Musk and Bryan Johnson have argued, necessitate this shift. In a world where humans are competing with artificial intelligence, the ancient motivator of "paying the bills" comes into play.

But there are other motivators at work.

Connecting our brains to the cloud provides us with a massive boost in processing power and memory, and, at least theoretically, can give us access to all the other minds online. Think of it this way: Computers, on their own, are interesting. But connect a bunch of those computers together, wire up a network, and you get the beginnings of the World Wide Web. Now imagine what happens when those computers are actually brains, which are already the most complicated machines in the known universe. And imagine that we don't just get to transmit thoughts, but feelings, experiences, and maybe, just maybe, meaning. If this were possible, would we hang on to our singular consciousness for long, or would we start to migrate into the collective mind that's evolving online?

Before you answer, consider three more details. First, we humans are an extremely social species. Loneliness, according to too many studies, is one of the great and deadly terrors of the modern era. The desire for connection is a foundational human driver, an intrinsic motivator in the psychological parlance. But it's not the only one in play.

The closest humans have come to a hive mind is the experience known as "group flow," the shared, collective version of a flow state. Group flow is a team performing at its very best: an incredible brainstorming session, a fantastic fourth-quarter comeback, a band coming together and blowing the roof off the auditorium. It's also considered the most pleasurable state on Earth. When psychologists ask people to rank their favorite experiences, group flow always tops that list. So the opportunity to have this experience essentially on demand will be a potent migratory driver as well.

Finally, there's evolution to consider.

Since the origin of life on this planet, the trajectory of evolution has always been from the individual to the collective. We went from single-celled organisms to multi-celled organisms to the massive multicellular organisms known as human beings. This is the typical thrust of natural selection, and why should today's selections be any different? There's little reason to believe that humanity has reached the pinnacle of intelligence, of development, of possibility, that reality television, that our megacities of tangled steel and endless asphalt, represent the best that life on Earth has to offer. Rather, we are merely a point on a spectrum, the ultimate "You are here" arrow.

There is, however, strong evidence that we won't be here for long. Neuralink has a plan for a two-gigabit-per-second wireless connection from the brain to the cloud and wants to begin human trials by the end of 2021. More and more—as this, and so many of the discoveries discussed in this book illustrate—the once slow and passive process of natural selection is being transformed into one rapid and proactive: evolution by human direction. This means, over the next century, technological acceleration may do more than just disrupt industries and institutions, it may actually disrupt the progress of biologically based intelligence on Earth. This break will birth a new species, one progressing at exponential speeds, both a mass migration and a meta-intelligence, and, ultimately, here at the tail end of our tale, yet another reason the future is faster than we think.

A meta-intelligence would be quite the innovation accelerant. If solitary minds working in collectivist organizations—aka, business, culture, and society—produced converging exponential technologies—aka, the fastest innovation accelerant the world has yet seen—imagine what a hive-minded planet—aka, a kinder, gentler

Borg—might be capable of creating. Put differently: How fast is our future if we're all thinking together?

And if you've come out the other side of all this thinking or feeling a little unsettled, there's actually a technical term for this as well: loss aversion. One of our most potent cognitive biases, loss aversion is the evolutionarily programmed suspicion that if I take away whatever you have today, whatever I replace it with tomorrow will be a whole lot worse. This is why people stay stuck in ruts, it's among the main reasons companies have such difficulty innovating, and why cultural change is so molasses slow.

Yet, who knows, maybe our hive mind will get us past this particular blind spot, but, until then, that sense of converging exponentials meets five great migrations meets holy-shit vertigo that you might be feeling is perfectly natural. As is the trepidation, the excitement, the unlocking of imagination. We feel it too. And all we can really say is what we've been saying to each other along the way: Take a deep breath and don't blink, because, ready or not, here comes tomorrow.

AFTERWORD

Abundance Revisited

One way to view the acceleration of technology described in this book is as part of a continuous march toward abundance. This was a theme we first introduced in our 2012 book, appropriately named: *Abundance: The Future Is Better Than You Think*. Since publication, this trend has only continued. There is no doubt that the cost of an ever-increasing number of goods and services has all but disappeared. Nor is there much doubt about the positive downstream effects of demonetization. Abundant cheap energy leads to abundant clean water. Autonomous electric cars lead to cheaper, greener transportation options and lower-cost access to housing. The combination of AI, 5G and AR/VR will provide low-cost education, entertainment, and healthcare to nearly every human on Earth, independent of geography or socioeconomic status.

There are, of course, plenty of reasons to disagree with this idea. The gap between the wealthy and poor grows ever wider, and the notion that there's an easy solution tucked inside our technology has been criticized as techno-utopian. But exponential technologies continue to march onward, and with them the ongoing process of demonetization and democratization.

In January 2019, for example, a *Wall Street Journal* headline reported: "The world is quietly getting better." The story that followed examined the latest World Bank numbers, which showed a continued decline in the number of individuals living below $2 per day, aka extreme poverty. And while it is true that the rich are getting richer, the poor are becoming much more enabled, with ever-increasing access to tools and technologies not measured by today's economic systems.

These two scenarios nearly always follow each other. The first mobile phones of the 1980s were slow and glitchy and served only the wealthy few. Today, however, while our phones are fast, flawless, and chock-full of capabilities, they are now cheap enough to be in the hands of even the poorest people on Earth. So, while we will see a future in which the rich may be living on Mars and accessing the latest in longevity treatments, this goes hand-in-hand with a future where everyone on Earth has increasingly low-cost access to food, energy, water, education, healthcare, and entertainment.

Our point is that, as *Abundance* reaches its tenth anniversary (2022), this concept is no longer simply a concept. Certainly, there's still a long way to go. A great many existing solutions are not yet globally distributed, and critical issues like water scarcity, climate change, and global hunger are heading in the wrong direction. Yet, as the *Wall Street Journal* pointed out, dozens of other indicators are trending up. In *The Better Angels of Our Nature*, to offer a different example, Harvard's Steven Pinker eloquently demonstrates that war and strife have reached all-time lows, and we are living in the most peaceful time in human history. Also the healthiest. Whether you're measuring by decreasing infant mortality and teenage birthrates, the number of deaths from malaria, the death toll from famine, or our ever-expanding life spans, here too the indicators show incredible progress. Meanwhile, the cost of unsubsidized renewable energy continues to plummet, while both high-speed digital connectivity and the availability of cheap, capable devices are both exploding. And with these devices and connectivity in place, a world of possibility has begun materializing before our eyes.

Today, a child in Tanzania has access to AI-enabled education tech, as well as the sum total of the world's available information via Google or Baidu. That same child, connected to the bandwidth explosion heading her way, will soon be able to spin up thousands of processor cores belonging to any number of cloud-based services, and tap into everything from the billions of hours of free entertainment on YouTube to our ever-thriving gig economy. Conveniently, the poorest nations on Earth are also the sunniest and with that sun—and the ever-increasing spread of solar—comes the opportunity for abundant energy. With energy comes the power to provide clean water and with clean water comes massive increases in health and wellness, which together with increased education and lower birth rates, can help stem the tide of overpopulation.

To be clear, there will still be terrorism, war, and murder. Dictatorship and disease won't go away. But the world will quietly continue to get better. And, as we described in *Abundance*, the goal here isn't about creating a life of luxury, but rather a life of possibility. Thanks to the forces of convergence, the technological advances needed for that world of abundance are coming at an ever-increasing pace. Of course, creating that world won't happen automatically. It will still require the largest cooperative effort in history. And this brings us to our final question: What, exactly, are you waiting for?

Which Way Next?

If you're tired of waiting, if the concepts and capabilities outlined in the book are of further interest, whether you're a CEO trying to steer your company through accelerating technological change or a garage-based entrepreneur looking to capitalize on these same changes, we've got you covered. Now, for certain, trying to grope with a century's worth of technological change unfolding over the next decade is a tall order, and trying to do this with our local and linear brains makes it ever more complicated.

The only way that your authors have found to navigate these waters, at least today, is through constant and continuous education.

We feel there are two critical components to this education, one mental, one physical. On the mental side, the significant performance upgrades that come from learning how to tap into the state of consciousness known as flow—amplified productivity, learning, creativity, collaboration, cooperation (this list goes on)—gives us the ability to keep pace. Flow augments all the brain's foundational information processing machinery, giving us the ability to think at speed and at scale—which are the two key cognitive requirements for thriving in an exponential world.

Simultaneously, there's a physical side to this equation—built out of actual, physical technology. Today's exponential entrepreneur and leader must diligently and continuously update their understanding of what technology now exists, and what that technology makes possible, moment to moment. While this kind of continuous learning is possible, it's not easy, which is perhaps one of the reasons for the increased popularity of Singularity University, Abundance360, and Abundance Digital—programs that constantly refresh participants on the state of the "exponentially possible."

Below, for both sides of this equation, you'll find a great list of options and opportunities.

Zero-to-Dangerous: The World's Leading Flow-Hacking Training: Zero to Dangerous is a research-backed peak-performance training designed to help entrepreneurs and leaders access the peak performance state known as flow. It is specifically focused on teaching people how to perform at speed and think at scale—which are the two keys to thriving in an exponential world. Zero to Dangerous blends cutting-edge science with one-on-one peak-performance coaching from licensed clinical psychologists and provides access to a network of high-performing individuals from across the globe. The training is led by Steven Kotler and leverages the same tools that he has used to train Google, US Navy SEALs, and Accenture. To learn more go to:

ZerotoDangerous.com. To learn more about Flow Research Collective, go to FlowResearchCollective.com.

Abundance360: A360 is a year-long membership (curated and run by Peter Diamandis) designed to help entrepreneurs navigate exponential technological change. Each year kicks off with a three-day mastermind in Beverly Hills, where members gather to learn about the latest breakthroughs in AI, networks, robotics, 3-D printing, AR/VR, biotech, and blockchain, and how these technologies are immediately applicable in their businesses and lives. Our mission is to provide members with the information, insights, and implementation tools they need to stay ahead of the curve.

Understanding how to navigate accelerating technology is essential for every entrepreneur. A360's mission is to give you the insights and tools to multiply by 10x your team's abilities, and to connect with other abundance and exponentially minded leaders. To apply for A360 go to www.A360.com.

Abundance Digital: Abundance Digital is the digitized, demonetized, and democratized version of Abundance360. The program services more than 2,000 entrepreneurs around the world at a 10x lower cost. The Abundance Digital community is your one-stop-shop for collaborating with other abundance and exponentially minded doers from around the world—including Peter Diamandis himself. Other benefits include:

- Dailey Coaching and Insights from the Abundance Digital APP and 100-plus hours of video content, teachings, and coaching by Dr. Diamandis.
- Livestream access to Peter's annual 3-Day Abundance360 Executive Summit, as well as Singularity University's Global Summit and Exponential Medicine program.
- Monthly Webinars: Live access to ~4 monthly video webinars featuring leading CEOs and entrepreneurs.

To apply go to www.Abundance.Digital

Singularity University: Singularity offers programs and events around the world focused on exponential-related teaching for graduate students, executives, and leaders. For a detailed list of the program offerings, go to www.SU.org.

XPRIZE Foundation: The XPRIZE Foundation uses large-scale global incentive competitions to crowdsource solutions to the world's grand challenges. XPRIZE believes that solutions can come from anyone, anywhere. Scientists, engineers, academics, entrepreneurs, and other innovators with new ideas from all over the world are invited to form teams and compete to win the prize. Rather than throw money at a problem, we incentivize the solution and challenge the world to solve it. The XPRIZE has designed and launched more than $200 million in prize purses across a range of subjects from space and oceans to education, food, water, energy, and environment just to name a few. To learn more about getting involved and/or competing go to www.XPRIZE.org.

Bold Capital Partners: Bold Capital Partners (BCP) manages a family of funds targeting investments in early-stage and growth technology companies, many of which are mentioned in this book. BCP is particularly interested in entrepreneurial leaders who leverage exponential technologies to transform the world and create innovative solutions to humanity's grand challenges. BCP was born from the convergence of three key forces: a leading vision of the future; a unique and valuable network of experts and practitioners supporting that vision; and a seasoned team of investment professionals capable of executing investments around that vision. Bold was named after Peter's second book (by the same name). For more information on the fund and the BCP portfolio of companies, go to: www.BoldCapitalPartners.com.

Human Longevity Inc.'s HEALTH NUCLEUS: In this book, we mention the potential breakthroughs racing our way in the field of longevity. One of the most important tools is made possible by HLI's

Health Nucleus program, which uses the latest technological advancements to provide a new standard for personal health. By integrating full-body imaging with whole-genome sequencing, Health Nucleus CORE is the first personalized assessment to reveal a more complete picture of your past, present, and future health status. You benefit from our highly experienced team of board certified and licensed physicians, geneticists, and expert scientists who use innovative technology to bring about the next big shift in quality of life. For more information, go to www.HumanLongevity.com.

Keynotes: Hiring Peter Diamandis and/or Steven Kotler

Both Peter and Steven enjoy speaking about their work in *Abundance*, *BOLD*, and *The Future Is Faster Than You Think.* They both do a limited number of select keynote speaking engagements ever year.

For more information on hiring Peter Diamandis, go to:
www.diamandis.com/speaking

For more information on hiring Steven Kotler, go to:
www.stevenkotler.com/speaking

ACKNOWLEDGMENTS

The Future Is Faster Than You Think has benefited greatly from the generous wisdom of a great many people. First off, the authors would like to express deep gratitude to their families—Jet, Dax, and Kristen Diamandis, and Joy Nicholson—for their incredible love, patience, and support. We'd also like to thank our agent, John Brockman; our editor; Stephanie Frerich, and everyone at Simon & Schuster who worked so doggedly on this project.

Also on the editing front, the always incredible Michael Wharton has been there every step of the way, and we owe him huge thanks for his insight, feedback, and incredible stamina. Gratitude to Max Goldberg for the staggering task of helping track down and organize our references, and to Jarom Longhurst for the book's flawless marketing campaign.

Special thanks also goes to Steven's team at the Flow Research Collective (especially Rian Doris) and Peter's team at PHD Ventures (Esther Count, Claire Adair, Max Goldberg, Derek Dolin, Kelley Lujan, Jarom Longhurst, Bri Lempesis, Greg O'Brien, Tom Compere, Sue Glanzrock, Joe Mosely, and Connie Fox) for their incredible support in doing research, crowdsourcing content, and blog creation, and for providing 24-7 support. And last but not least, special recognition to Esther Count and Connie Fox for the herculean task of coordinating Peter's schedule and life.

On the research and inspiration front, we are grateful to the Singularity University family of alumni, faculty, and staff under the leadership of cofounder and chancellor Ray Kurzweil, associate founder Rob Nail, chief growth officer Carin Watson, and executive chairman Erik Anderson. Thank you also to the XPRIZE family under the leadership of Anousheh Ansari for the offices we occupied and the inspiring stories of innovation we were able to tell in this book.

Finally, Peter wants to say thank-you to Dan Sullivan and the Strategic Coach team for the encouragement, wisdom and support on creating a 10x impact on the world.

NOTES

PART ONE: THE POWER OF CONVERGENCE

Chapter One: Convergence

Flying Cars

3 *In 2018, for the sixth straight year:* Inrix, "Global Traffic Scorecard," available here: http://inrix.com/scorecard/.

3 *ground zero for Uber Elevate:* See: https://www.uber.com/us/en/elevate/summit/2018/.

4 *"We've come to accept":* You can find Holden's original speech at https://www.youtube.com/watch?v=fmW2Y2nEW1U&feature=youtu.be.

4 *In the late 1990s:* Sarah Perez, "Groupon Product Chief Jess Holden to Depart, Is Heading to a Bay Area Tech Company," *TechCrunch*, February 11, 2014. See: https://techcrunch.com/2014/02/11/groupon-product-chief-jeff-holden-departs-is-headed-to-a-bay-area-tech-company/.

4 *100 million Prime members later:* Dennis Green, "A Survey Found That Amazon Prime Membership Is Soaring to New Heights—But One Trend Should Worry the Company," *Business Insider*, January 18, 2019, p. 1.

4 *a series of unlikely wins:* For Holden's full bio, check out his Linkedin: https://www.linkedin.com/in/jeffholden/.

5 *By mid-2019, over $1 billion:* Tim Fernholz, "Are There Bubbles in Space," *Quartz*, July 30, 2018. See: https://qz.com/1343920/investors-have-pumped-nearly-1-billion-into-aerospace-start-ups-this-year/.

5 *Larry Page:* Mark Harris, "Larry Page Is Quietly Amassing a 'Flying Car' Empire," *Verge*, July 19, 2018. See: https://www.theverge.com/2018/7/19/17586878/larry-page-flying-car-opener-kitty-hawk-cora.

5 *59 cents per passenger mile:* AAA, "AAA Reveals True Cost of Vehicle Ownership," August 23, 2017. See: https://newsroom.aaa.com/tag/driving-cost-per-mile/.

5 *For comparison, a helicopter:* This comparison was made by Uber, as part of their internal feasibility study. See: https://www.cnet.com/roadshow/news/will-you-be-able-to-afford-uberairs-flying-car-service/.

5 *For its 2020 launch:* Ibid.

6 *eVTOLs are being developed:* Ibid.

6 *They've also teamed up with:* For a full breakdown of Uber's partners, see: https://www.uber.com/us/en/elevate/partners/.

6 *Vehicles capable of flight:* "Vimana" is the name of the mythological flying chariots describe in early Hindu texts. See: https://en.wikipedia.org/wiki/Vimana.

7 *Even the more modern incarnations:* Steven Kotler, *Tomorrowland* (New Harvest, 2015), pp. 97–105.

Converging Technology

7 *Moore's Law:* See: https://www.intel.com/content/www/us/en/silicon-innovations/moores-law-technology.html.

8 *as a human brain:* Ray Kurzweil, *How to Create a Mind* (Viking, 2012), pp. 179–198.

8 *"Law of Accelerating Returns":* Ray Kurzweil, "The Law of Accelerating Returns," March 7, 2001. See: https://www.kurzweilai.net/the-law-of-accelerating-returns.

9 *we use the term "disruptive innovation":* Clayton Christensen, *The Innovator's Dilemma* (HarperBusiness, 2000), pp. 15–19.

10 *Enter distributed electric propulsion, or DEP for short:* Mark Moore, "Distributed Electric Propulsion Aircraft," Nasa Langley Research Center. See: https://aero.larc.nasa.gov/files/2012/11/Distributed-Electric-Propulsion-Aircraft.pdf.

10 *These electric engines are 95 percent:* Technically, the full range is between 90 percent and 98 percent, but for a breakdown and comparison to a gas motor, see: Karim Nice and Jonathon Strickland, "Gasoline and Battery Power Efficiency," *How Stuff Works,* https://auto.howstuffworks.com/fuel-efficiency/alternative-fuels/fuel-cell4.htm.

10 *To achieve this lift:* Holden interview, Ibid.

11 *still unbroken record:* Staff at Henry Ford, "Willow Run Bomber Plant." See: https://www.thehenryford.org/collections-and-research/digital-collections/expert-sets/101765/.

12 *Put it this way, in "The Law of Accelerating Returns,":* ibid.

More Transportation Options

13 *Model T:* History.com editors, "Ford Motor Company Unveils the Model T," *History*, August 27, 2009. See: https://www.history.com/this-day-in-history/ford-motor-company-unveils-the-model-t.

13 *Just four years later:* Elizabeth Kolbert, "Hosed," *New Yorker*, November 8, 2009.

13 *radio-controlled "American wonder":* Fabian Kroger, "Automated Driving in Its Social, Historical and Cultural Contexts," *Autonomous Driving*, May 22, 2016, pp. 41–68.

13 *Many date the pivotal breakthrough to 2004:* See DARPA's website for the full breakdown of events: https://www.darpa.mil/about-us/timeline/-grand-challenge-for-autonomous-vehicles.

13 *dozens of vehicles had logged millions:* Alexis Madrigal, "Waymo's Robots Drove More Miles Than Everyone Else Combined," *Atlantic*, February 14, 2009, See: https://www.theatlantic.com/technology/archive/2019/02/the-latest-self-driving-car-statistics-from-california/582763/.

14 *Waymo purchased that fleet:* Andrew Hawkins, "Waymo and Jaguar Will Build Up to 20,000 Self-Driving Electric SUVs," *Verge*, March 27, 2018. See: https://www.theverge.com/2018/3/27/17165992/waymo-jaguar-i-pace-self-driving-ny-auto-show-2018.

14 *To compete with Waymo, General Motors:* See GM's original press release: https://media.gm.com/media/us/en/gm/news.detail.html/content/Pages/news/us/en/2018/may/0531-gm-cruise.html.

14 *"Faster than anyone expects":* Author interview, 2019.

15 *80 percent cheaper than individual car ownership:* Ibid.

15 *average U.S. roundtrip commute:* U.S. Census Bureau, "Average One-Way Commuting Time by Metropolitan Areas," December 7, 2017. See: https://www.census.gov/library/visualizations/interactive/travel-time.html.

15 *there were a hundred plus automotive brands:* You can find an aggregated list of car brands, both in service and retired, at this Wikipedia page: https://en.wikipedia.org/wiki/List_of_car_brands.

15 *the average car owner:* Donald Shoup, *The High Cost of Free Parking* (Routledge, 2011), p. 624.

16 *America has almost half-a-million parking spaces:* Richard Florida, "Parking Has Eaten American Cities," *CityLab*, July 24, 2018.

16 *MIT professor of urban planning:* Eran Ben-Joseph, *ReThinking a Lot* (MIT Press, 2012), pp. xi–xix.

16 *Hyperloop is the brainchild:* For the original whitepaper: https://www.spacex.com/sites/spacex/files/hyperloop_alpha.pdf.

17 *Robert Goddard:* Malcolm Browne, "New Funds Fuel Magnet Power for Trains," *New York Times*, March 3, 1992.

17 *RAND corporation:* Robert Salter, "The Very High Speed Transit," Rand Corporation, 1972. See: https://www.rand.org/pubs/papers/P4874.html.

17 *In January 2013:* For the full story of Hyperloop One development, see: https://hyperloop-one.com/our-story#partner-program. (Author note: Peter's VC firm is an investor.)

17 *Josh Giegel*: Author interview, 2019.

18 *It also gave him time to tweet:* See: https://twitter.com/elonmusk.

19 *with $113 million of Musk's own money:* Dana Hull, "Musk's Boring Co. Raises $113 Million for Tunnels, Hyperloop," *Bloomberg*, April 16, 2018, See: https://www.bloomberg.com/news/articles/2018-04-16/musk-s-boring-co-raises-113-million-for-tunnels-and-hyperloop.

19 *a three-stop subway:* Aarian Marshall, "Las Vegas Orders Up a Boring Company Loop," *Wired*, May 22, 2019.

19 *electric boring machines:* Ed Oswald, "Here's Everything You Need to Know About the Boring Company," *Digital Trends*, February 26, 2019.

20 *the International Astronautical Congress in Adelaide, Australia:* For the full address, check out: https://www.youtube.com/watch?v=tdUX3ypDVwI.

20 *Musk announced his intentions to retire his current rocket fleet:* Darrell Etherington, "SpaceX aims to Replace Falcon 9, Falcon Heavy and Dragon with One Spaceship," *Techcrunch*, September 28. 2017. See: https://techcrunch.com/2017/09/28/spacex-aims-to-replace-falcon-9-falcon-heavy-and-dragon-with-one-spaceship/.

20 *LA mayor Eric Garcetti:* See: https://twitter.com/mayorofla.

20 *the very first test flights:* See: https://spacenews.com/spacex-begins-starship-hopper-testing/.

21 *Sears was worth $14.3 billion:* Each of the data points (company, year, and market cap) was taken from https://www.macrotrends.net. We use peak market cap values for the given year.

21 *Target $38.2*: Ibid.

21 *Walmart a whopping $158 billion:* Ibid.

21 *Amazon was at $17.5 billion:* Ibid.

21 *Hard times hit Main Street:* For all of the company data points in this paragraph, likewise refer to: Ibid.

21 *Studies done with fMRI:* For example, see: Arnaud D'Argembeau, "Modulation of Medial Prefrontal and Inferior Parietal Cortices When Thinking About Past, Present and Future Selves," *Social Neuroscience*, May 2, 2010, pp. 187–200.

22 *The convergence of AI and robotics could threaten:* For an overview of most of the major studies, see: Jill Lepore, "Are Robots Competing for Your Job?,"

New Yorker, March 4, 2019. For a different overview: Marguerite Ward, "AI and Robots Could Threaten Your Career Within 5 Years," CNBC, October 5, 2017, https://www.cnbc.com/2017/10/05/report-ai-and-robots-could-change-your-career-within-5-years.html.

23 *McKinsey Global Research:* Matthieu Pelissie due Rausas, "Internet Matters: The Net's Sweeping Impact on Growth, Jobs, and Prosperity," McKinsey Global Institute, May 2011.

23 *Yale's Richard Foster:* Richard Foster and Sarah Kaplan, *Creative Destruction* (Crown Business, 2001). As this original research was conducted with Innosight, see their executive summary for a quick overview: https://www.innosight.com/insight/creative-destruction/.

26 *In 2018, All Nippon Airways:* For the official announcement, see: https://avatar.xprize.org/prizes/avatar.

Chapter Two: The Jump to Lightspeed: Exponential Technologies, Part One

Quantum Computing

27 *The coldest place in the universe:* Author interview with Chad Rigetti, 2018.

27 *Back in 1995, astronomers in Chile:* Public Information Office, Jet Propulsion Laboratory, "Boomerang Nebula Boasts Coolest Spot in the Universe," June 20, 1997. For the official NASA/JPL release, see: https://www.jpl.nasa.gov/news/releases/97/coldspot.html.

28 *IBM's Deep Blue:* Luke Harding and Leonard Barden, "Deep Blue Win a Giant Step for Computerkind," *Guardian*, May 12, 2011.

28 *Transistor power:* Erik Brynjolfsson and Andrew McAfee, *The Second Machine Age* (W.W. Norton and Co., 2014), p. 49.

29 *Moore's Law has been slowing down:* Lieven Eeckhout, "Is Moore's Law Slowing Down? What Next?, *IEEE Micro* 37, no. 4: 4–5.

29 *"Moore's Law was not the first":* Kurzweil, "Law of Accelerating Returns."

29 *Apple's recent A12 Bionic:* See: https://www.apple.com/iphone-xs/a12-bionic/.

30 *Rose's Law:* Tim Ferriss does a good job overviewing this idea and its history here: https://tim.blog/2018/05/31/steve-jurvetson/.

30 *a fifty-quibit computer:* Rigetti, author interview.

30 *Oxford's Simon Benjamin:* This came from a talk he gave at the Oxford Martin School, February 2016. See: https://www.oxfordmartin.ox.ac.uk/videos/the-dawn-of-quantum-technology-with-prof-simon-benjamin/.

30 *over 120 million programs:* Rigetti, author interview.

32 *twenty-six websites online:* See: https://www.internetlivestats.com/total-number-of-websites/.

Artificial Intelligence

33 *Her name was Xiaoice:* Geoff Spencer, "Much More Than a Chat: China's Xiaoice Mixes AI with Emotions and Wins Over Millions of Fans," news.microsoft.com, November 1, 2018. See: https://news.microsoft.com/apac/features/much-more-than-a-chatbot-chinas-xiaoice-mixes-ai-with-emotions-and-wins-over-millions-of-fans/. See also: https://blogs.microsoft.com/ai/xiaoice-full-duplex/.

33 *"Yeah, magnets":* This and the below example about my girlfriend being mad at me are taken from author interviews with Zo, the American version of Xiaoice that was released on Twitter in 2018.

33 *Dragon TV:* Matt McFarland, "What Happened When a Chinese TV Station Replaced Its Meteorologist with a Chatbot," *Washington Post*, January 12, 2016.

34 *consider the service economy:* John Ward, "The Services Sector: How Best to Measure It?," *International Trade Organization*, October 2010.

34 *forty-three different types of traffic signs:* To track the progress in machine learning, Wikipedia has a useful chart here: https://en.wikipedia.org/wiki/Timeline_of_machine_learning. See also: Andrew McAfee and Erik Brynjolfsson, *Machine Platform Crowd* (Norton, 2017), pp. 66–86.

35 *AI-piloted drone:* For a demo, see: https://www.youtube.com/watch?v=gsfkGlSajHQ.

35 *AI assistant named Duplex:* See: https://ai.googleblog.com/2018/05/duplex-ai-system-for-natural-conversation.html.

35 *Google's Talk to Books:* See: https://experiments.withgoogle.com/talk-to-books.

35 *Japan's national literary:* Chloe Olewitz, "A Japanese AI Program Just Wrote a Short Novel, and It Almost Won a Literary Prize," *Digital Trends*, March 23, 2016.

36 *Google's AlphaGo:* For a great breakdown of the difference between chess and Go, see: Danielle Muoio, "Why Go Is So Much Harder for AI to Beat Than Chess," *Business Insider*, March 10, 2016. For information on Lee Sodol's defeat, see also: Jon Russell, "Google AI Beat Go World Champion Again to Complete Historic 4–1 Series," *Techcrunch*, March 15, 2016.

36 *an AI build another AI:* Aatif Sulleyman, "Google AI Creates Its Own Child AI That's More Advanced Than Systems Built by Humans," *Independent*, December 5, 2017.

36 *Facebook relies on AI:* Megan Dickey, "Facebook Brings Suicide Prevention Tools to Live and Messenger," *TechCrunch*, March 1, 2017.

36 *the Department of Defense:* Steven Kotler and Jamie Wheal, *Stealing Fire* (Harper Collins, 2018), pp. 100–102.

37 *AI ran for mayor:* Paul Withers, "Robots Take Over," *Express*, April 17, 2018. We choose this references because it has the best video. See: https://www.express.co.uk/news/world/947448/robots-japan-tokyo-mayor-artificial-intelligence-ai-news.

Networks

38 *Change came on May 24, 1844:* Library of Congress staff, "Invention of the Telegraph," Samuel F. B. Morse Papers at the Library of Congress, 1793–1919. See: https://www.loc.gov/collections/samuel-morse-papers/articles-and-essays/invention-of-the-telegraph/.

38 *Alexander Graham Bell:* See: https://www.loc.gov/item/today-in-history/march-10/.

38 *In 1919, less than 10 percent of U.S. households*: For a great history of mass communications development, see: https://www.elon.edu/e-web/predictions/150/1870.xhtml.

38 *Today it's around 28 cents:* See: https://www.verizon.com/personal/info/international-calling/.

39 *In 2010, around a quarter of the Earth's population:* The World Bank keeps good numbers here: https://data.worldbank.org/indicator/SP.POP.TOTL.

39 *By 2017, penetration had reached:* Ibid.

39 *network evolution:* G. Smilarubavathy, "The Survey on Evolution of Wireless Network Generations," *International Journal of Science, Technology and Engineering* 3, no. 5 (November 2016).

39 *Project Loony:* See: https://loon.com/.

40 *launching OneWeb:* Sarah Scoles, "Maybe Nobody Wants Your Space Internet," *Wired*, March 15, 2018.

40 *Amazon joined this satellite competition:* Alan Boyle, "Amazon to Offer Broadband Access from Orbit," *GeekWire*, April 4, 2019. See: https://www.geekwire.com/2019/amazon-project-kuiper-broadband-satellite/.

40 *SpaceX:* For the FCC's original press release, see: https://www.fcc.gov/document/fcc-authorizes-spacex-provide-broadband-satellite-services.

41 *In 2014, at an infectious disease lab in Finland:* Author interview with Oura CEO Harpreet Singh, 2018. (Author note: Peter's VC firm is an investor.)

42 *the Oura ring:* See: http://ouraring.com. (Author note: Peter's VC firm is an investor.)

42 *John Romkey:* Ryan Nagelhout, *Smart Machines and the Internet of Things* (Rosen Publishing, 2016).

42 *Neil Gross*: Neil Gross, "The Earth Will Don an Electronic Skin," *BusinessWeek*, August 29, 1999.

43 *In 2009, the number of devices connected to the Internet:* Dave Evans, "The Internet of Things," Cisco.com, April 2011. See: https://www.cisco.com/c/dam/en_us/about/ac79/docs/innov/IoT_IBSG_0411FINAL.pdf.

43 *By 2015, all this progress added up to 15 billion:* Louis Columbus, "Roundup of Internet of Things Forecasts and Market Estimates, 2016," Forbes.com, no. 27 (2016). See: https://www.forbes.com/sites/louiscolumbus/2016/11/27/roundup-of-internet-of-things-forecasts-and-market-estimates-2016/#32a1a5ba292d.

43 *Nor will we stop there*: For a full breakdown of the Accenture report, see: https://newsroom.accenture.com/subjects/management-consulting/industrial-internet-of-things-will-boost-economic-growth-but-greater-government-and-business-action-needed-to-fulfill-its-potential-finds-accenture.htm.

43 *Steven Sasson:* Steven Kotler and Peter Diamandis, *BOLD* (Simon & Schuster, 2015), pp 4–6.

43 *LIDAR sensors:* Sean Higgins, "Livox Announces $600 Lidar for Autonomous Vehicles," Spar3-D.com, January 23, 2019. See: https://www.spar3-D.com/news/lidar/livox-announces-600-lidar-for-autonomous-vehicles-uav-mapping-and-more-an-its-shipping-now/.

43 *The first commercial GPS:* For a great graphic breakdown of the full history of GPS, see: https://onlinemasters.ohio.edu/blog/the-evolution-of-portable-gps/. See also: https://web.mit.edu/digitalapollo/Documents/Chapter6/hoagprogreport.pdf.

44 *smart dust:* Brendan Koerner, "What Is Smart Dust Anyway," *Wired*, June 1, 2003. See also: https://news.berkeley.edu/2018/04/10/berkeley-engineers-build-smallest-volume-most-efficient-wireless-nerve-stimulator/.

44 *The data haul from these sensors:* For a great overview, see: https://newsroom.intel.com/news/2018-ces-keynote-intel-brian-krzanich/#gs.yzs68u.

Robotics

45 *In March 2011, an earthquake in Tokyo*: For the full International Atomic Energy Commission report on Fukushima, see: https://www-pub.iaea.org/MTCD/Publications/PDF/Pub1710-ReportByTheDG-Web.pdf.

45 *The disaster hit Honda:* Evan Ackerman, "Honda Halts Asimo Development in Favor of More Useful Humanoid Robots," *IEEE Spectrum*, June 28, 2018. Also try: https://blogs.wsj.com/japanrealtime/2011/04/20/the-little-robot-that-couldnt/.

45 *DARPA launched their Robotics Challenge:* Katie Drummond and Noah Shachtman, "Darpa's Next Grand Challenge," *Wired*, April 5, 2012.

46 *Gill Pratt:* For Gill Pratt's quote on the Challenge, see: https://spectrum.ieee.org/automaton/robotics/humanoids/darpa-robotics-challenge-amazing-moments-lessons-learned-whats-next.

46 *Boston Dynamics' robot Atlas:* See: https://www.bostondynamics.com/atlas.

46 *Honda also got in on the action:* Evan Ackerman, "Honda Unveils Prototype E2-D2 Disaster Response Robot," IEEE Spectrum, October 2, 2017.

46 *Softbank:* Ingrid Lunden, "Softbank Is Buying Robotics Firm Boston Dynamics and Schaft from Alphabet," *TechCrunch*, June 8, 2017.

46 *After decades of rising life expectancies*: See: https://www.economist.com/graphic-detail/2019/07/09/japans-pension-problems-are-a-harbinger-of-challenges-elsewhere. See also: this article in the *Japan Times*: https://www.japantimes.co.jp/news/2019/06/04/business/financial-markets/japans-pension-system-inadequate-aging-society-council-warns/#.XWayvi2ZOWY.

47 *robo-mason:* Check out the video: https://www.youtube.com/watch?v=HiOkXKb1DBk.

47 *The UR3:* See: https://www.universal-robots.com/products/ur3-robot/.

47 *Amazon's also been driving the drone segment*: See: https://www.amazon.com/Amazon-Prime-Air.

48 *Hurricane Sandy:* Edward Baig, "Cell Service Can Mean Life or Death After a Disaster. Can Drones Help?," *USA Today*, March 16, 2018. See also: https://www.nytimes.com/2018/04/06/nyregion/drone-cellphone-disaster-service.html.

48 *Boeing's heavy-lifting drones:* Alex Davies, "Boeing's Experimental Cargo Drone Is a Heavy Lifter," *Wired*, January 14, 2018.

48 *Zipline:* See: https://flyzipline.com.

48 *tree-planting drones:* https://www.biocarbonengineering.com.

Chapter Three: The Turbo-Boost: Exponential Technologies, Part II

Virtual and Augmented Reality

49 *Jeremy Bailenson:* Author interview, 2019.

50 *the concept of VR has been around since the sixties:* Patent - US2955156A, "Stereoscopic-Television Apparatus for Individual Use."

50 *you could purchase the EyePhone:* Simon Parkin, "Virtual Reality Startups Look Back to the Future,"March 7, 2014, *Technology Review*. See: https://www.technologyreview.com/s/525301/virtual-reality-startups-look-back-to-the-future. See also: "A Whole New Universe," *New York*, August 6, 1990, p. 32.

51 *$2 billion to acquire the VR company Oculus Rift:* "Facebook to Acquire Oculus," March 25, 2015.Facebook Newsroom, See: https://newsroom.fb.com/news/2014/03/facebook-to-acquire-oculus/.

51 *By 2015, Venture Beat reported that a market which typically saw only ten*

new entrants a year, suddenly had 23: Dean Takahashi, "The Landscape of VR Is Complicated—with 234 Companies Valued at $13B," *Venture Beat*, October 12, 2015.

51 *The year 2017 was a banner one for Samsung, when they sold 3.65 million headsets:* Shanhong Liu, "Worldwide Virtual Reality (VR) Headset Unit Sales by Brand in 2016 and 2017 (In Millions)," *Statista*, August 9, 2019. See: https://www.statista.com/statistics/752110/global-vr-headset-sales-by-brand/.

51 *Apple:* Jeremy Horwitz, "Apple Lists AR/VR Jobs, Reportedly Taps Executive Who Finalizes Products," *Venture Beat*, August 1, 2019.

51 *Google:* See: https://vr.google.com/.

51 *Cisco:* Jens Meggers, "Virtual Reality, Meet Cisco Spark," September 18, 2017, Cisco.com. See: https://blogs.cisco.com/collaboration/cisco-spark-in-virtual-reality.

51 *Microsoft:* Adi Robertson, "Microsoft Says It's No Longer Planning VR Support on Xbox," *Verge*, June 20, 2018. See: https://www.theverge.com/2018/6/20/17485852/microsoft-xbox-one-no-vr-headset-support-windows-mixed-reality-e3-2018.

51 *Phone-based VR showed up soon afterward:* Google Cardboard. See: https://vr.google.com/cardboard/.

51 *By 2018, the first wireless adaptors, standalone headsets, and mobile headsets had hit the market:* "Introducing Oculus Quest, Our First 6DOF All-In-One VR System, Launching Spring 2019," September 28, 2018,Oculus.com, See https://www.oculus.com/blog/introducing-oculus-quest-our-first-6dof-all-in-one-vr-system-launching-spring-2019/?locale=en_US. See also: "HTC VIVE Unveils VIVE Focus Plus Pricing, Availability, Improved Connectivity, and Enhanced Lenses," March 25, 2019, HTC.com, https://www.htc.com/us/newsroom/2019-03-25/.

51 *Google and LG doubled:* Carlin Vieri, "An 18 Megapixel 4.3,Ä≥ 1443 Ppi 120 Hz OLED Display for Wide Field of View High Acuity Head Mounted Displays," *Society for Information Display*, May 9, 2018, pp. 314–324. See also: Stefan Etienne, "Google and LG Show Off Their High-Res VR Display for Future Headsets," *Circuit Breaker*, May 23, 2018, https://www.theverge.com/circuitbreaker/2018/5/23/17383990/google-lg-vr-display-high-res-headsets.

51 *HEAR360's "omni-binaural" microphone suite captures 360 degrees of audio*: Hear360. See: https://hear360.io/#8ball.

51 *Touch has also reached the masses:* Sarah Needleman, "Virtual Reality, Now with the Sense of Touch," *Wall Street Journal*, April 3, 2018.

51 *Scent emitters, taste simulators:* For example, see: the Feelreal sensory mask: https://feelreal.com/.

51 *brainwave readers:* For example, see: Neurable: http://www.neurable.com/.

51 *eMarketer study:* Victoria Petrook, "Virtual and Augmented Reality Users 2019," *eMarketer*, March 27, 2019. See: https://www.emarketer.com/content/virtual-and-augmented-reality-users-2019. See also: "Forecast for the Number of Active Virtual Reality Users Worldwide from 2014 to 2018 (in Millions)," Statista, https://www.statista.com/statistics/426469/active-virtual-reality-users-worldwide/.

51 *VR market around $35 billion or so:* "Profiles in Innovation: Virtual and Augmented Reality," January 13, 2016.GoldmanSachs.com, See: https://www.goldmansachs.com/insights/pages/technology-driving-innovation-folder/virtual-and-augmented-reality/report.pdf.

52 *He's developed first-person VR experiences:* Author interview with Bailenson. See also: Jeremy Bailenson, *Experience on Demand: What Virtual Reality Is, How It Works, and What It Can Do* (W. W. Norton & Company, 2018).

52 *2010 speech at NYU law school:* "The Virtues of Virtual Reality: How Immersive Technology Can Reduce Bias," April 26, 2019 (video). See: https://www.youtube.com/watch?v=vXxfkkINq8M.

52 *In 2016, when Nintendo's Pokemon GO was downloaded almost a billion times:* Lauren Musni, "Pokémon GO Surpasses the 1 Billion Downloads Milestone," *Nintendo Wire*, July 31, 2019.

52 *coming out with an AR developers suite that lets anyone design apps for their platform:* You can find details about Apple's AR development suite here: https://developer.apple.com/augmented-reality/.

52 *purchasing Akonia Holographics:* Lucas Matney, "Apple Buys Denver Startup Building Waveguide Lenses for AR Glasses," *TechCrunch*, August 29, 2018.

52 *eighteen hundred different AR startups:* You can run your own search with this URL: https://angel.co/companies?markets[]=Augmented+Reality.

52 *a market in excess of $133 billion:* "Global Augmented Reality (Ar) Market Will Reach USD 133.78 Billion by 2021," Zion Market Research, November 24, 2016.

52 *$100 will get you an entry-level Leap Motion headset:* Jeremy Horwitz, "Leap Motion Shows Crazy-Looking $100 North Star AR Headset with Hand Tracking," *Venture Beat*, April 9, 2018.

52 *$3000 covers a top-shelf Microsoft HoloLens:* Mariella Moon, "Microsoft HoloLens 2 Will Go On Sale in September (Update)," *Engadget*, August 29, 2019. Learn more about the Hololens here: https://www.microsoft.com/en-us/hololens.

3-D Printing

53 *The most expensive supply chain in the universe extends only 241 miles:* Remy Melina, "International Space Station: By the Numbers," Space.com, August

4, 2017. See: https://www.space.com/8876-international-space-station-numbers.html.

53 *It costs $10,000 per pound to get an object out of the Earth's gravity well:* "Advanced Space Transportation Program: Paving the Highway to Space," National Aeronautics and Space Administration (NASA), April 12, 2018. See: https://www.nasa.gov/centers/marshall/news/background/facts/astp.html. See also: Robert Dempsey, "The International Space Station Operating an Outpost in the New Frontier," National Aeronautics and Space Administration (NASA), April 13, 2018, https://www.nasa.gov/sites/default/files/atoms/files/iss-operating_an_outpost-tagged.pdf.

53 *a significant portion of ISS's precious real estate is taken up by the storage of replacement parts:* Ibid.

53 *Made In Space:* See: https://madeinspace.us/.

53 *Well, it's a few years later and Made In Space is now in space:* Quincy Bean, "3-D Printing in Zero-G Technology Demonstration," National Aeronautics and Space Administration (NASA). See: https://www.nasa.gov/mission_pages/station/research/experiments/explorer/Investigation.html?#id=1039.

53 *Which is why, on a 2018 ISS mission, when an astronaut broke his finger:* JY Wong, "On-Site 3-D Printing of Functional Custom Mallet Splints for Mars Analogue Crewmembers," *Aerospace Medicine and Human Performance*, October 2015, doi: 10.3357/AMHP.4259.2015, pp. 911–914. See also: "3-D Printing the First Medical Supplies on the Space Station," January 12, 2017, http://www.3-D4md.com/blog/2017/1/12/3-D-printing-the-first-medical-supplies-on-the-space-station.

53 *The original 3-D printers showed up back in the eighties:* Dana Goldberg, *Autodesk.com*, "History of 3-D Printing: It's Older Than You Are (That Is, If You're Under 30)," April 13, 2013. See: https://www.autodesk.com/redshift/history-of-3-D-printing/.

54 *hundreds of different materials:* Author interview with Avi Reichental, CEO of Exponential Works, 2018.

54 *jet engines:* Matthew Van Dusen, "GE's 3-D-Printed Airplane Engine Will Run This Year. General Electric," June 19, 2017. See: https://www.ge.com/reports/mad-props-3-D-printed-airplane-engine-will-run-year/.

54 *apartment complexes:* Brittney Sevenson, "Shanghai-Based WinSun 3-D Prints 6-Story Apartment Building and an Incredible Home," 3-DPrint.com, January 8, 2015. See: https://3-Dprint.com/38144/3-D-printed-apartment-building/.

54 *circuit boards:* Nano Dimension Inc.. See: https://www.nano-di.com/.

54 *prosthetic limbs:* For a database with examples of 3-D printed prosthetics,

see*:* "3-D-Printable Prosthetic Devices," National Institutes of Health, https://3-Dprint.nih.gov/collections/prosthetics.

54 *$12 trillion manufacturing sector:* Eric Gjovik, "Additive Manufacturing and Its Impact on a $12 Trillion Industry," May 14, 2019. See: https://www.manufacturing.net/2019/05/additive-manufacturing-and-its-impact-12-trillion-industry.

54 *Until the early 2000s, 3-D printers were exceptionally pricey:* Blake Griffin, "New Report Shows Manufacturing Output Hit $35 Trillion in 2017," Interact Analysis. See: https://www.interactanalysis.com/new-report-shows-manufacturing-output-hit-35-trillion-in-2017-growth-forecast-to-continue/.

54 *for under $1000:* B.T. Wittbrodta, "Life-Cycle Economic Analysis of Distributed Manufacturing with Open-Source 3-D Printers," *Mechatronics,* September 2013, 713–726. See: https://www.sciencedirect.com/science/article/pii/S0957415813001153.

54 *performance has increased*: Avi Reichental, author interview.

54 *Nano Dimension converged:* Lucas Mearin, "3-D Printer Presages the Future of Multi-Layer Circuit Board Design," *ComputerWorld.* See: https://www.computerworld.com/article/3195839/desktop-3-D-printer-presages-the-future-of-multi-layer-circuit-board-design.html.

54 *batteries:* "3-D-Printed Lithium-Ion Batteries," American Chemical Society, October 17, 2018. See: https://www.acs.org/content/acs/en/pressroom/presspacs/2018/acs-presspac-october-17-2018/3-D-printed-lithium-ion-batteries.html.

54 *wind turbines:* Jon Fingas, "3-D-Printed Wind Turbine Puts 300W of Power in Your Backpack," *Engadget*, August 17, 2014. See: https://www.engadget.com/2014/08/17/airenergy-3-D-wind-turbine/. See also: "Transforming Wind Turbine Blade Mold Manufacturing with 3-D Printing," https://www.energy.gov/eere/wind/videos/transforming-wind-turbine-blade-mold-manufacturing-3-D-printing.

54 *solar cells:* Santanu Bag, "Aerosol-Jet-Assisted Thin-Film Growth of CH3NH3PbI3 Perovskites—A Means to Achieve High Quality, Defect-Free Films for Efficient Solar Cells," *Advanced Energy Materials*, July 14, 2017. See also: Corey Clark, "Air Force Research Laboratory Creates 3-D Printed Solar Cells," *3-D Printing Industry*, July 19, 2017.

54 *GE's advanced turboprop:* Tomas Kellner, "Fired Up: GE Successfully Tested Its Advanced Turboprop Engine with 3-D-Printed Parts," *General Electric*, January 2, 2018. See: https://www.ge.com/reports/ge-fired-its-3-D-printed-advanced-turboprop-engine/.

54 *The first 3-D printed prosthetics arrived in 2010:* See this history of 3-D printing: https://3-Dinsider.com/3-D-printing-history/.

54 *In 2018, a Jordanian hospital:* Hanna Watkin, "Doctors Without Borders Hospital in Jordan 3-D Print Prostheses for War Victims," *All About 3-D Printing*, December 10, 2018. See: https://all3-Dp.com/4/doctors-without-borders-hospital-jordan-3-D-print-prostheses-war-victims/.

55 *Unlimited Tomorrow:* See: https://www.unlimitedtomorrow.com/product/.

55 *Open Bionics:* Retrieved from https://openbionics.com/.

55 *3-D-printed multi-grip bionic prosthetics at non-bionic prices:* Jelle ten Kate, "3-D-Printed Upper Limb Prostheses: A Review," *Assistive Technology*, February 2, 2017, pp. 300–314.

55 *first kidney tissue capable of filtering blood and producing urine:* Anthony Atala, "Printing a Human Kidney," TED, 2018. See: https://www.ted.com/talks/anthony_atala_printing_a_human_kidney?language=en. See also: Kate Yandell, "Organs on Demand," *Scientist*, September 1, 2013, https://www.the-scientist.com/features/organs-on-demand-38787. See also: Patent - US6673339B1, "Prosthetic Kidney and Its Use for Treating Kidney Disease."

55 *In 2010, Organovo, a San Diego-based bioprinting outfit:* Vanesa Listek, "Organovo: Bioprinting Could Be the New Solution to Organ Transplantation," 3-DPrint.com, August 27, 2019. See also: Kena Hudson, "First Fully Bioprinted Blood Vessels," *Business Wire*, December 8, 2010, https://www.businesswire.com/news/home/20101208006587/en/Fully-Bioprinted-Blood-Vessels. See also: Karoly Jakab, "Tissue Engineering by Self-Assembly and Bio-Printing of Living Cells," IOP Science, June 2, 2010, https://iopscience.iop.org/article/10.1088/1758-5082/2/2/022001.

55 *Prellis Biologics:* See: https://www.prellisbio.com/. See also: Scott Claire, "Prellis Biologics Reaches Record Speed and Resolution in Viable 3-D Printed Human Tissue," 3-DPrint.com, https://3-Dprint.com/217267/prellis-biologics-record-speed/.

55 *Iviva Medical is doing the same with 3-D printed kidneys:* Author conversation with Iviva Medical CEO Dr. Brock Reeve. See also: https://ivivamedical.com/. (Author note: Peter's VC firm is an investor.)

55 *3-D printed organs are predicted to hit the market by 2023:* Ibid.

55 *WinSun 3-D printed ten single-family homes in under twenty-four hours:* "3-D Printers Print Ten Houses in 24 Hours" (video), April 16, 2014. See: https://www.youtube.com/watch?v=SObzNdyRTBs. See also: "China: Firm 3-D Prints 10 Full-Sized Houses in a Day," *BBC News*, April 25, 2014, https://www.bbc.com/news/blogs-news-from-elsewhere-27156775.

55 *each costing less than $5000:* Leo Gregurić, "How Much Does a 3-D Printed

House Cost in 2019?," February 12, 2019. See: https://all3-Dp.com/2/3-D -printed-house-cost/.

55 *In 2017, a different Chinese company:* "Chinese Construction Firm Erects 57 -Storey Skyscraper in 19 Days," *Guardian*, April 30, 2015.

55 *In 2019, the California-based Mighty Building:* Author interview, 2019. (Author note: Peter's VC firm is an investor.)

55 *Brett Hagler:* Author interview, 2018. For more background on Brett Hagler's company, New Story, see: Adele Peters, "There Will Soon Be a Whole Community of Ultra-Low-Cost 3-D-Printed Homes," *Fast Company*, March 11, 2019, https://www.fastcompany.com/90317441 /there-will-soon-be-a-whole-community-made-of-these-ultra-low-cost-3-D -printed-homes.

56 *In the fall of 2019, in Mexico:* Ibid.

Blockchain

56 *were first proposed in 1983:* David Chaum, "Blind Signatures for Untraceable Payments," *Advances in Cryptography* (Springer 1998), pp. 199–203 See: http://blog.koehntopp.de/uploads/Chaum. BlindSigForPayment.1982.PDF.

57 *Satoshi Nakamoto:* Satoshi Nakamotoe, "Bitcoin: A Peer-to-Peer Electronic Cash System." See: https://bitcoin.org/bitcoin.pdf.

57 *In 2010, Laszlo Hanyecz solved that problem:* Nick Bilton, "Disruptions: Betting on a Coin with no Realm," *New York Times*, December 22, 2013.

57 *By 2019, they were just shy of $15,000:* Data retreived from: https:// coinmarketcap.com/currencies/bitcoin/.

57 *$308 billion:* "Billion Reasons to Bank Inclusively." See: https://www .accenture.com/us-en/_acnmedia/accenture/conversion-assets/dotcom /documents/global/pdf/dualpub_22/accenture-billion-reasons-bank -inclusively.pdf#zoom=50.

58 is *worth $600 billion:* "Mitigation and Remittances," World Bank Group, 2018. See: https://www.knomad.org/sites/default/files/2018-04 /Migration%20and%20Development%20Brief%2029.pdf.

58 *All of that money is being skimmed off the top:* Katie Lobosco, "Walmart Offers Less Costly Money Wire Service," CNN, April 17, 2014. See: https://money.cnn.com/2014/04/17/news/companies/walmart-money -transfers/index.html. You can also find Western Union's Fee Table here: https://www.westernunion.com/content/dam/wu/EU/EN /feeTableRetailEN-ES.PDF.

58 *they also lack an official identity:* Desai Vyjayanti, "The Global Identification Challenge: Who Are the 1 Billion People Without Proof of Identity?"

World Bank, April 25, 2018. See: https://blogs.worldbank.org/voices/global-identification-challenge-who-are-1-billion-people-without-proof-identity.

58 *peer-to-peer ridesharing:* Paul Vigna, *The Truth Machine: The Blockchain and the Future of Everything* (Macmillan Publishing Group, 2018), p. 7.

58 *validate any asset:* Elizabeth Paton, "Will Blockchain Be a Boon to the Jewelry Industry?," *New York Times*, November 30, 2018.

58 *Sports betting is one example:* Gerald Fenech, "Blockchain in Gambling and Betting: Are There Real Advantages?," *Forbes*, January 30, 2019.

59 *J.P. Morgan, Goldman Sachs, and Bank of America:* Goldman: Alastair Marsh, "Goldman Sachs Explores Creating a Digital Coin like JPMorgan's," *Bloomberg*, June 28, 2019. J.P. Morgan: Hugh Son, "JP Morgan Is Rolling Out the First US Bank-Backed Cryptocurrency to Transform Payments Business," CNBC, February 14, 2019. Bank of America: Hugh Son, "Bank of America Tech Chief Is Skeptical on Blockchain Even Though BofA Has the Most Patents for It, CNBC, March 26, 2019.

59 *a market value of almost $10 billion as of 2018:* "Funds Raised in 2018." See https://www.icodata.io/stats/2018.

59 *grow to $176 billion by 2025, and could exceed $3.1 trillion in 2030:* "Gartner Predicts 90% of Current Enterprise Blockchain Platform Implementations Will Require Replacement by 2021." See: https://www.gartner.com/en/newsroom/press-releases/2019-07-03-gartner-predicts-90--of-current-enterprise-blockchain.

59 *Vatom, Inc.:* See: https://www.crunchbase.com/organization/vatomic.

Material Science and Nanotechnology

61 "*materials science" problem:* "History of the Light Bulb," Department of Energy, November 22, 2013. See: https://www.energy.gov/articles/history-light-bulb.

61 *fourteen months testing over sixteen hundred materials:* Joyce Bedi, "Thomas Edison's Inventive Life. Lemelson Center," April 18, 2004. See https://invention.si.edu/thomas-edisons-inventive-life.

61 *creating a bulb capable of twelve hundred hours:* "Edison Files," http://edisonmuseum.org/content3399.html.

61 *Much brighter and longer lasting tungsten filaments:* "Incandescent Lamp with Ductile Tungsten Filament." Americanhistory.edu, See: https://americanhistory.si.edu/collections/search/object/nmah_704238.

62 *Materials Genome Initiative:* President Barack Obama, "Remarks by the President at Carnegie Mellon University's National Robotics Engineering Center," Office of the Press Secretary: The White House, June 24, 2011. See also: "The First Five Years of the Materials Genome Initiative:

Accomplishments and Technical Highlights," https://mgi.gov/sites/default/files/documents/mgi-accomplishments-at-5-years-august-2016.pdf.

62 *Jeff Carbeck, the head of Advanced:* Author interview, 2018.

63 *carbon fiber composites for lighter-weight vehicles, advanced alloys for more durable jet engines:* Adrian P. Mouritz, "Introduction to Aerospace Materials," *Introduction to Aerospace Materials* (Woodhead Publishing Limited, 2012), pp. 1–14.

63 *biomaterials to replace human joints:* Kalpana S Katti, "Biomaterials in Total Joint Replacement," *Colloids and Surfaces B: Biointerfaces,* 2004, pp. 133–142.

63 *energy storage:* Yayuan Liu, "Design of Complex Nanomaterials for Energy Storage: Past Success and Future Opportunity," *Accounts of Chemical Research*, December 5, 2017, pp. 2895–2905,

63 *quantum computing:* "Nanotechnology for Quantum Computers, Industry Skills for Physics Students, Technologies That Make Physics Happen," *Physics World*, August 1, 2019. See: https://physicsworld.com/a/nanotechnology-for-quantum-computers-industry-skills-for-physics-students-technologies-that-make-physics-happen/.

63 *Omkaram Nalamasu:* Author interview, 2019.

63 *Right now, the "conversion efficiency":* Ran Fu, "U.S. Solar Photovoltaic System Cost Benchmark: Q1 2018," *National Renewable Energy Laboratory*, 2018. See: https://www.nrel.gov/docs/fy19osti/72399.pdf.

63 *Perovskite:* Brian Wang, "First Commercial Perovskite Solar Late in 2019 and the Road to Moving the Energy Needle," *Next Big Future*, February 3, 2019. See: https://www.nextbigfuture.com/2019/02/first-commercial-perovskite-solar-late-in-2019-and-the-road-to-moving-the-energy-needle.html.

63 *Richard Feynman's 1959 speech, "There's Plenty of Room at the Bottom":* Richard P. Feynman, "There's Plenty of Room at the Bottom," *Engineering and Science*, 1960.

63 *K. Eric Drexler's 1987 book:* Eric Drexler, *Engines of Creation: The Coming Era of Nanotechnology (Anchor Library of Science)* (Anchor Books, 1987).

64 *Researchers at Harvard built a nanoscale 3-D printer:* Dan Ferber, "Printing Tiny Batteries, Wyss Institute," June 18, 2013. See: https://www.seas.harvard.edu/news/2013/06/printing-tiny-batteries.

64 *smart contact lenses with a resolution six times greater:* Author conversation with Steve Sinclair, SVP, Mojo Vision, 2018. (Author note: Peter's VC firm is an investor.)

64 *drug delivery nanobots:* Suping Li, "A DNA Nanorobot Functions as a Cancer Therapeutic in Response to a Molecular Trigger in Vivo," *Nature Biotechnology*, 2018, pp. 258–264.

64 *seven hundred terabytes of data in a single gram of DNA:* Megan Molteni, "The Rise of DNA Data Storage," *Wire*, 2018. See: https://www.wired.com/story/the-rise-of-dna-data-storage/. More recently, "Catalog Successfully Stores All 16GB of Wikipedia Text on DNA," *Verdict*, July 9, 2019. See: https://www.verdict.co.uk/dna-data-storage-2019/.

64 *a system 10 percent the size of the Sahara Desert:* Author conversation with Bill Gross, CEO, Idealab, 2018.

Biotechnology

65 *The 1970s were good for John Travolta:* "John Travolta," IMDb. See: https://www.imdb.com/name/nm0000237/.

65 *1975 TV show* Welcome Back, Kotter*:* "Welcome Back, Kotter," IMDb. See: https://www.imdb.com/title/tt0072582/.

65 *made-for-TV movie* The Boy in the Plastic Bubble*:* "The Boy in the Plastic Bubble," IMDb. See: https://www.imdb.com/title/tt0074236/.

65 *an article in* Science *argued:* Theodore Friedmann, "Gene Therapy for Human Genetic Disease?," *Science*, March 1972, pp. 949–955. See: https://science.sciencemag.org/content/175/4025/949.long.

65 *In 1999, an eighteen-year-old boy named Jesse Gelsinger:* Sheryl Gay Stolberg, "The Biotech Death of Jesse Gelsinger," *New York Times Managzine*, November 28, 1999. See: https://www.nytimes.com/1999/11/28/magazine/the-biotech-death-of-jesse-gelsinger.html.

66 *in a gene therapy trial in France aimed at treating Bubble Boy disease:* "Why Gene Therapy Caused Leukemia in Some 'Boy in the Bubble Syndrome' Patients," *Journal of Clinical Investigation*, August 10, 2008. See: https://www.sciencedaily.com/releases/2008/08/080807175438.htm.

66 *Bubble Boy disease had been cured:* "Gene Therapy Cures Babies with 'Bubble Boy' Disease," *Genetic Engineering & Biotechnoogy News,* August 19, 2019. See: https://www.genengnews.com/topics/gene-therapy-cures-babies-with-bubble-boy-disease/.

66 *With over fifty gene therapy drugs in the final phases of clinical trials:* "Gene Therapy Phase 4." See: https://clinicaltrials.gov/ct2/results?term=gene+therapy&age_v=&gndr=&type=&rslt=&phase=3&Search=Apply.

66 *human body, which is a collection of 30 to 40 trillion cells:* Rose Eveleth, "There Are 37.2 Trillion Cells in Your Body," *Smithsonian Magazine*, October 24, 2013. See: https://www.smithsonianmag.com/smart-news/there-are-372-trillion-cells-in-your-body-4941473/.

67 *This was the goal of the Human Genome Project:* "Humane Genome Results." See: https://www.genome.gov/human-genome-project/results.

67 *Since then, though, the price has plummeted:* "DNA Sequencing Costs: Data."

Genome.gov, See: https://www.genome.gov/about-genomics/fact-sheets/DNA-Sequencing-Costs-Data.

67 *Today, sequencing a human genome takes a few days:* Ibid.

67 *CRISPR-Cas9, for example:* "CRISPR 2.0: Genome Engineering Made Easy as A-B-C," November 5, 2017. Hardvard.edu, See: http://sitn.hms.harvard.edu/flash/2017/crispr-2-0-genome-engineering-made-easy-b-c/.

67 *scientists at Harvard unveiled CRISPR 2.0:* Eric S. Lander, "The Heroes of CRISPR," *Cell,* January 14, 2016, pp. 18–28. See: https://www.cell.com/cell/fulltext/S0092-8674(15)01705-5?_returnURL=https%3A%2F%2Flinkinghub.elsevier.com%2Fretrieve%2Fpii%2FS0092867415017055%3Fshowall%3-Dtrue.

67 *David Liu, the Harvard chemical biologist who led the work, told the* LA Times: Deborah Netburn, "New Gene-Editing Technique May Lead to Treatment for Thousands of Diseases," *LA Times,* October 25, 2017. See: https://www.latimes.com/science/sciencenow/la-sci-sn-dna-gene-editing-20171025-story.html.

67 *Human germline engineering is another CRISPR application:* Antonion Regalado, "EXCLUSIVE: Chinese Scientists Are Creating CRISPR Babies," *MIT Review,* November 25, 2018. See: https://www.technologyreview.com/s/612458/exclusive-chinese-scientists-are-creating-crispr-babies/. See also: Heidi Ledford, "CRISPR Babies: When Will the World Be Ready?," *Nature,* June 19, 2019, https://www.nature.com/articles/d41586-019-01906-z.

68 *There's also stem cells to consider:* "Stem Cell Information," National Institutes of Health. See: https://stemcells.nih.gov/info/basics/1.htm.

Chapter Four: The Acceleration of Acceleration

Force #1: Saved Time

70 *In "The Original Macintosh":* Andy Hertzfeld, "Saving Lives," *Folklore.com,* August 1983. See: https://www.folklore.org/StoryView.py?story=Saving_Lives.txt.

71 *University of Michigan behavioral economist Yan Chen:* Yan Chen, "A Day Without a Search Engine: An Experimental Study of Online and Offline Searches," *Experimental Economics* 17, no. 4 (December 2014): 512–536. See: https://link.springer.com/article/10.1007/s10683-013-9381-9.

71 *Over the past hundred years, labor-saving devices:* University of Montreal, "Fridges and Washing Machines Liberated Women, Study Suggests," *Science Daily,* March 13, 2009. See: https://www.sciencedaily.com/releases/2009/03

/090312150735.htm. Original paper can be found here: https://pdfs.semanticscholar.org/423-D/28062802774c5687bd2545c4024a4961085e.pdf.

72 *New York to Chicago was four weeks by stagecoach:* "Maps of the Day: Travel Times from NYC in 1800, 1830, 1857 and 1930." AEI.org, See: http://www.aei.org/publication/maps-of-the-day-travel-times-from-nyc-in-1800-1830-1857-and-1930.

Force #2: Availability of Capital

73 *Sputnik 1 into orbit:* Tim Wallace, "How Sputnik 1 Launched the Space Age," *Cosmos Magazine*, October 4, 2017. See: https://cosmosmagazine.com/space/how-sputnik-1-launched-the-space-age.

73 *Edward Teller:* Ibid.

73 *Senator Mike Mansfield:* Paul Dickson, *Sputnik* (Walker, 2001), p. 116.

73 *Yuri Gagarin:* Ibid., p. 215.

73 *pouring 2.2 percent of the US GDP into the aerospace industry:* Deborah D. Stine, "The Manhattan Project, the Apollo Program, and Federal Energy Technology R&D Programs: A Comparative Analysis," *Congressional Research Service*, June 30, 2009. See: https://fas.org/sgp/crs/misc/RL34645.pdf.

74 *The very first crowdfunding project:* Mike Mettler, "Prog Legends Marillion Have Mastered Crowdfunding, High-Res Rock," *Digital Trends*, December 2, 2016. See: https://www.digitaltrends.com/music/interview-mark-kelley-of-marillion/.

74 *by 2015, a worldwide total of $34 billion:* Ben Paynter, "How Will the Rise of Crowdfunding Reshape How We Give to Charity?" *Fast Company*, March 3, 2017, See: https://www.fastcompany.com/3068534/how-will-the-rise-of-crowdfunding-reshape-how-we-give-to-charity-2.

74 *Kickstarter:* See: https://www.kickstarter.com/help/stats.

74 *Pebble Time:* John McDermott, "Pebble 'Smartwatch' Funding Soars on Kickstarter," *Inc.*, April 20, 2012. See: https://www.inc.com/john-mcdermott/pebble-smartwatch-funding-sets-kickstarter-record.html.

74 *$300 billion:* Massolution/Crowdsourcing.org, 2015 CF Crowdfunding Industry Report. See: http://reports.crowdsourcing.org/index.php?route=product/product&product_id=54.

75 *crowdfunding as "potentially the most disruptive of all the new models of finance.":* The Future of Finance, the Socialization of Finance (Goldman Sachs Report, March 2015). See: https://www.planet-fintech.com/downloads/The-future-of-Finance-the-Socialization-of-Finance-Golman-Sachs-march-2015_t18796.html.

75 *$8.1 billion in 1995 to:* See: https://www.nsf.gov/statistics/issuebrf/sib99303.htm#nsb.

75 *$61.4 billion in 2016: PwC | CB Insights MoneyTree™ Report Q4 2018*, p. 6, found here: https://www.pwc.com/us/en/moneytree-report/moneytree-report-q4-2018.pdf.

75 *investments reached $99.5 billion:* Ibid., p. 10.

75 *Asia, a relatively new player, peaked at $48 billion:* Ibid, p. 85.

75 *European VCs set an all-time high: $21 billion:* Ibid, p. 82.

75 *$5.4 billion in 2017 to $9.3 billion in 2018:* Ibid, p. 20.

75 *$11.8 billion in 2017 to $14.4 billion in 2018:* 3Q 2018 PitchBook-NVCA Venture Monitor Report, found here: https://files.pitchbook.com/website/files/pdf/3Q_2018_PitchBook_NVCA_Venture_Monitor.pdf.

76 *Filecoin:* For this and all below references to biggest ICOs, please see Oscar Williams-Grut, "The 11 Biggest ICO Fundraises of 2017," *Business Insider*, January 1, 2018. See: https://www.businessinsider.com/the-10-biggest-ico-fundraises-of-2017-2017-12.

76 *$8.5* trillion *in assets:* Data aggregated from this wikipedia article: https://en.wikipedia.org/wiki/Sovereign_wealth_fund, by way of the Sovereign Wealth Fund Institute database at https://www.swfinstitute.org.

76 *forty-two SWF deals valued at around $16.2 billion:* Claire Milhench, "Sovereign Investors Hunt for 'Unicorns' in Silicon Valley," Reuters, May 11, 2017.

76 *"I totally believe [in] this concept,":* Sam Shead, "The Japanese Tech Billionaire Behind Softbank Thinks the 'Singularity' Will Occur Within 30 Years," *Business Insider*, February 27, 2017. See: https://www.businessinsider.com/softbank-ceo-masayoshi-son-thinks-singularity-will-occur-within-30-years-2017-2.

77 *The Vision Fund got started:* "Masayoshi Son Prepares to Unleash His Second $100bn Tech Fund," *Economist*, March 23, 2019. See: https://www.economist.com/business/2019/03/23/masayoshi-son-prepares-to-unleash-his-second-100bn-tech-fund.

Force #3: Demonetization

78 *Ilumina's latest generation sequencer:* Sarah Buhr, "Illumina Wants to Sequence Your Whole Genome for $100," *TechCrunch*, January 10, 2017. See: https://techcrunch.com/2017/01/10/illumina-wants-to-sequence-your-whole-genome-for-100/.

78 *a 2019 report by the International Renewable Energy Agency:* See full IRENA press release: https://www.irena.org/newsroom/pressreleases/2019/Apr/Renewable-Energy-Now-Accounts-for-a-Third-of-Global-Power-Capacity.

Force #4: More Genius

79 *It was signed S. Ramanujan:* Robert Kanigel, *The Man Who Knew Infinity* (Washington Square Press, 1992).

80 *the standard distribution of the Stanford-Binet scale:* For a good overview, see Duke University intelligence researcher Jonathon Wai's piece for *Psychology Today*: https://www.psychologytoday.com/us/articles/201207/brainiacs-and-billionaires.

81 *In a study run at the University of Sydney:* Richard Chi and Allan Snyder, "Brain Stimulation Enables the Solution of an Inherently Difficult Problem," *Neuroscience Letters* 515, no. 2 (May 2, 2012): 121–124.

81 *Neuralink:* See: https://neuralink.com.

81 *Kernel:* See: https://kernel.co.

81 *as Johnson explains:* Author interview with Bryan Johnson, 2018.

82 *2017, USC neuroscientist Doug Song:* Eileen Toh, "USC Researchers Develop Brain Implant to Improve Memory," *USC Daily Troject*, November 19, 2017. See: http://dailytrojan.com/2017/11/19/usc-researchers-develop-brain-implant-improve-memory/.

82 *the full cyborg to the middle 2030s:* Jillian Eugenios, "Ray Kurzweil: Humans Will Be Hybrids by 2030," CNN, June 4, 2015. See: http://money.cnn.com/2015/06/03/technology/ray-kurzweil-predictions/.

82 *86 percent success rate:* Dominic Basulto, "Why Ray Kurzweil's Predictions Are Right 86% of the Time," *Big Think*, December 13, 2012. See: https://bigthink.com/endless-innovation/why-ray-kurzweils-predictions-are-right-86-of-the-time.

Force #5: Communications Abundance

82 *as author Matt Ridley:* Matt Ridley, *Rational Optimist* (HarperCollins, 2010), p. 1.

82 *Santa Fe Institute physicist Geoffrey West:* West wrote a great piece on all this work for *Medium*. Find it here: https://medium.com/sfi-30-foundations-frontiers/scaling-the-surprising-mathematics-of-life-and-civilization-49ee18640a8.

83 *roughly one-quarter of the Earth's population, some 1.8 billion people, were connected to the internet. By 2017, that penetration had reached 3.8 billion people:* "Individuals Using the Internet (% of population)," World Bank. See: https://data.worldbank.org/indicator/IT.net.USER.ZS. See also: "Population, Total," World Bank, https://data.worldbank.org/indicator/SP.POP.TOTL.

Force #6: New Business Models

83 *a 2015 article in the* McKinsey Quarterly*:* Marc de Jong, "Disrupting Beliefs: A New Approach to Business-Model Innovation," *McKinsey*

Quarterly, July 2015. See: https://www.mckinsey.com/business-functions/strategy-and-corporate-finance/our-insights/disrupting-beliefs-a-new-approach-to-business-model-innovation.

83 *"bait and hook":* Randal C. Picker, "The Razors-and-Blades Myth(s)," *University of Chicago Law Review*, February 6, 2011. See: https://lawreview.uchicago.edu/publication/razors-and-blades-myths.

83 *"franchise models":* Kerry Pipes, "History of Franchising: Franchising in the Modern Age," Franchising.com. See: https://www.franchising.com/guides/history_of_franchising_part_two.html.

85 *In his excellent book* The Inevitable*:* Kevin Kelly, *The Inevitable* (Viking, 2016), p. 33.

85 *Decentralized Autonomous Organizations:* Christoph Jentzsch, "Decentralized Orgnizations to Automate Government." See: https://archive.org/stream/DecentralizedAutonomousOrganizations/WhitePaper_djvu.txt.

86 *the multimillion-dollar economy:* Maria Korolov, "Second Life GDP Totals $500 Million," *Hypergrid Business*, November 11, 2015. See: https://www.hypergridbusiness.com/2015/11/second-life-gdp-totals-500-million/.

86 *The Experience Economy*: Joseph Pine and James Gilmore, "Welcome to the Experience Economy," *Harvard Business Review*, July-August 1998.

87 *Harvard's Clayton Christensen:* Mark Johnson, *Reinvent Your Business Model* (Harvard Business Press, 2018), back cover.

Force #7: Longer Lives

87 *Ada Lovelace:* Walter Isaacson, *The Innovators* (Simon & Schuster, 2014) pp. 7–33.

89 *the average caveperson hit puberty:* For a great review of lifespan through history, see: https://ourworldindata.org/life-expectancy.

89 *cardiac disease and cancer:* Author interview with Robert Hariri, MD, PhD, 2018.

89 *"Google's New Project to Solve Death":* Harry McCracken, "How CEO Larry Page Has Transformed the Search Giant into a Factory for Moonshots. Our Exclusive Look at His Boldest Bet Yet—to Extend Human Life," *TIME*, September 30, 2013.

89 *"Google Wants to Cheat Death":* Connor Simpson, "Google Wants to Cheat Death," *Atlantic*, September 18, 2013.

90 *Jeff Bezos–backed Unity Biotechnology:* See: https://unitybiotechnology.com.

90 *middle-aged mice these same medicines:* Jan M. van Deursen, "Senolytic Therapies for Healthy Longevity," *Science* 364, no. 6441: 636–637.

90 *Back in 2014, Stanford researchers:* Saul A Villeda, "Young Blood Reverses

Age-Related Impairments in Cognitive Function and Synaptic Plasticity in Mice," *Nature Medicine* 20 (2014): 659–663.

90 *Elevian, for example, a Harvard University spinoff:* See: https://www.elevian.com. (Author note: Peter's VC firm is an investor.)

90 *When injected into older mice:* **For hearts:** Francesco Loffredo, "Growth Differentiation Factor 11 Is a Circulating Factor That Reverses Age-Related Cardiac Hypertrophy," *Cell* 153, no. 4 (May 9, 2013): 828–239. **For brains:** Lida Katsimpardi, "Vascular and Neurogenic Rejuvenation of the Aging Mouse Brain by Young Systemic Factors," *Science* 344, no, 6184 (May 9, 2014): 630–634. **For muscles:** Manisha Sinha, "Restoring Systemic GDF11 Levels Reverses Age-Related Dysfunction in Mouse Skeletal Muscle," *Science* 344, no. 6184 (May 9, 2014): 649–652. **For lungs:** Katsuhiro Onodera, "Decrease in an Anti-Ageing Factor, Growth Differentiation Factor 11, in Chronic Obstructive Pulmonary Disease," *Thorax* 72, no. 10 (April 28, 2017). **For kidneys:** Y. Zhang, "GDF11 Improves Tubular Regeneration After Acute Kidney Injury in Elderly Mice," *Nature Scientific Reports* 6 (October 5, 2016).

90 *Samumed, LLC, for example:* Ibid.

90 *$13 billion valuation:* Lydia Ramsey, "Samumed, a $12 Billion Startup That Wants to Cure Baldness and Smooth Out Your Wrinkles, Just Raised Even More Funding as It Plots an IPO," *Business Insider*, August 11, 2018. See: https://www.businessinsider.com/samumed-raises-438-million-at-12-billion-valuation-2018-8.

90 *Celularity:* See: https://www.celularity.com.

90 *placental-derived stem cells can extend life 30 to 40 percent:* Hariri, author interview.

91 *"longevity escape velocity":* Ray Kurzweil, author interview, 2018. For a video, see: https://singularityhub.com/2017/11/10/3-dangerous-ideas-from-ray-kurzweil/.

PART TWO: THE REBIRTH OF EVERYTHING

Chapter Five: The Future of Shopping

The First Platform Play

95 *Richard Warren Sears was born on December 7, 1863:* Vicki Howard, "The Rise and Fall of Sears," *Smithsonian Magazine*, July 25, 2017. See: https://www.smithsonianmag.com/history/rise-and-fall-sears-180964181/. See also: "Richard Warren Sears: Biography & Sears, Roebuck, & Company,"

https://schoolworkhelper.net/richard-warren-sears-biography-sears-roebuck-company/, and "Richard W. Sears," https://www.britannica.com/biography/Richard-W-Sears.

96 *Back then, every city tracked time by its own clock:* "Why Do We Have Time Zones?" TimeandDate.com, See: https://www.timeanddate.com/time/time-zones-history.html.

97 *But in 1896, Congress passed the Rural Free Delivery Act:* "Rural Free Delivery," August 2013. See: https://about.usps.com/who-we-are/postal-history/rural-free-delivery.pdf.

97 *"Within 10 years," wrote journalist Derek Thompson in the* Atlantic*:* Derek Thompson, "Sears Is Not a Failure," *Atlantic*, October 15, 2018. See: https://www.theatlantic.com/ideas/archive/2018/10/end-sears/573070/.

97 *Author Jeremy Rifkin points out that all major economic paradigm shifts share a common denominator:* Elena Holodny, "A Key Player in China and the EU's 'Third Industrial Revolution' Describes the Economy of Tomorrow," *Business Insider*, July 16, 2017. See: https://www.businessinsider.com/jeremy-rifkin-interview-2017-6.

98 *After a 132-year run, in the autumn of 2018, Sears filed for bankruptcy:* Thompson, "Sears Is Not a Failure."

98 *Amazon, a company founded about the same time Walmart was disrupting Sears*: Marisa Gertz, "How One of America's Oldest Retailers Unraveled," *Bloomberg*, October 12, 2018. See: https://www.bloomberg.com/news/photo-essays/2018-10-12/how-sears-got-left-behind-as-walmart-amazon-took-over-retail. See also: Matt Day, "The Enormous Numbers Behind Amazon's Market Reach," *Bloomberg*, March 27, 2019. https://www.bloomberg.com/graphics/2019-amazon-reach-across-markets/.

99 *in 2017 alone, the total was 6,700:* "Global Powers of Retailing 2018," Deloitte. See: https://www2.deloitte.com/content/dam/Deloitte/tr/Documents/consumer-business/cip-2018-global-powers-retailing.pdf. See also: Jackie Wattles, "2017 Just Set the All-Time Record for Store Closings," CNN, October 25, 2017, https://money.cnn.com/2017/10/25/news/economy/store-closings-2017/index.html?sr=twCNN102517economy0528PMStory.

99 *All one needs to do is look at the following table:* Data from https://www.macrotrends.net; we report the peak value for each company during the listed year.

99 *Even though online sales increased from $34 billion in Q1 of 2009 to $115 billion in Q3 of 2017:* Census.com, "Monthly Retail Trade." See: https://www.census.gov/retail/index.html. See also: "Retail E-Commerce Sales in the United States from 1st Quarter 2009 to 2nd Quarter 2019 (In Million

U.S. Dollars)," Statista, https://www.statista.com/statistics/187443/quarterly-e-commerce-sales-in-the-the-us/.

99 *this growth spurt only accounted for 10 percent of total retail sales:* "Quarterly Retail E-Commerce Sales 3rd Quarter 2017," U.S. Census Bureau. See: https://www2.census.gov/retail/releases/historical/ecomm/17q3.pdf.

99 *3.8 billion in 2017:* "Individuals Using the Internet (% of population)," World Bank. See: https://data.worldbank.org/indicator/IT.net.USER.ZS. See also: "Population, Total," World Bank, https://data.worldbank.org/indicator/SP.POP.TOTL.

99 *8.2 billion in 2025:* Daniel Goodkind, "The World Population at 7 Billion," U.S. Census Bureau, October 31, 2011. See: https://www.census.gov/newsroom/blogs/random-samplings/2011/10/the-world-population-at-7-billion.html.

AI and the Retail Experience

100 *Nokia was the world leader in cell phones, but when smartphones showed up, they ended up out of business:* James Surowiecki, "Where Nokia Went Wrong," *New Yorker*, September 3, 2013. See: https://www.newyorker.com/business/currency/where-nokia-went-wrong. See also: Haydn Shaughnessy, "Apple's Rise and Nokia's Fall Highlight Platform Strategy Essentials," *Forbes*, March 8, 2013, https://www.forbes.com/sites/haydnshaughnessy/2013/03/08/apples-rise-and-nokias-fall-highlight-platform-strategy-essentials/#575a41346e9a.

101 *Captain Kirk talk to the* Enterprise*'s computer on* Star Trek*:* "Star Trek," IMDb. See: https://www.imdb.com/title/tt0060028/.

101 *$2 billion today to $8 billion by 2023:* "Digital Voice Assistants in Use to Triple to 8 Billion by 2023, Driven by Smart Home Devices," Juniper Research, February 12, 2018. See: https://www.juniperresearch.com/press/press-releases/digital-voice-assistants-in-use-to-8-million-2023.

101 *$1,700:* Eugene Kim, "Amazon Echo Owners Spend More on Amazon Than Prime Members, Report Says," CNBC, January 3, 2018. See: https://www.cnbc.com/2018/01/03/amazon-echo-owners-spend-more-on-amazon-than-prime-members.html.

101 *2018 demo of Google Duplex:* "Google Duplex: An AI System for Accomplishing Real-World Tasks over the Phone," Google AI Blog, May 8, 2018. See: https://ai.googleblog.com/2018/05/duplex-ai-system-for-natural-conversation.html.

101 *Google I/O conference, seven thousand developers:* Abner Li, "Googler in Charge of I/O 2019 Says It Takes 6-9 Months to Plan," *9To5Google*, May 6, 2019.

101 *the soft-spoken Google CEO Sundar Pichai may have stolen back the crown:*

"Keynote (Google I/O '18)." See: https://www.youtube.com/watch?v=ogfYd705cRs.

102 *According to a recent Zendesk study:* "The Impact of Customer Service on Customer Lifetime Value 2013." See: https://www.zendesk.com/resources/customer-service-and-lifetime-customer-value/.

102 *Beyond Verbal:* See: https://beyondverbal.com/.

102 *Based on research conducted on more than seventy thousand subjects in more than thirty different languages:* Ben Woods, "Emotion Analytics Company Beyond Verbal Releases Moodies as Standalone Ios App." TheNextWeb.com, See: https://thenextweb.com/apps/2014/01/23/beyond-verbal-releases-moodies-standalone-ios-app/.

103 *New Zealand's Soul Machines:* Greg Cross, author interview, 2018. See also: https://www.soulmachines.com/.

103 *Powered by IBM's Watson:* "Soul Machines." IBM.com, See: https://www.ibm.com/case-studies/soul-machines-hybrid-cloud-ai-chatbot.

103 *Software manufacturer Autodesk now includes a Soul Machine avatar named AVA:* Kari Johnson, "How Autodesk's Assistant Ava Attempts to Avoid Uncanny Valley," *Venture Beat*, May 18, 2018. See: https://venturebeat.com/2018/05/18/how-autodesks-assistant-ava-attempts-to-avoid-uncanny-valley/.

103 *For Daimler Financial Services:* "Hot Off the Press: Emotional Intelligence Daimler Financial Services Invests in Soul Machines," *Soul Machines*, October 17, 2018. See: https://www.soulmachines.com/news/2018/10/17/hot-off-the-press-emotional-intelligence-daimler-financial-services-invests-in-soul-machines/.

Go, Go, Gone are Cashiers

105 *By 2025, according to research done by McKinsey:* See: https://www.mckinsey.com/~/media/mckinsey/business%20functions/mckinsey%20digital/our%20insights/the%20internet%20of%20things%20the%20value%20of%20digitizing%20the%20physical%20world/unlocking_the_potential_of_the_internet_of_things_full_report.ashx.

105 *Automatic checkout:* "Beyond Amazon Go: The Technologies and Players Shaping Cashier-Less Retail," *CB Insights*, October 9, 2018. See: https://www.cbinsights.com/research/cashierless-retail-technologies-companies-trends/.

105 *the initial Go store opened for business in Seattle:* Nick Wingfield, "Inside Amazon Go, a Store of the Future," *New York Times*, January 21, 2018. See: https://www.nytimes.com/2018/01/21/technology/inside-amazon-go-a-store-of-the-future.html.

105 *The following year Go opened seven more stores and have plans for three*

thousand more by 2021: Brad Stone, "Amazon's Most Ambitious Research Project Is a Convenience Store," *Bloomberg Businessweek*, July 18, 2019. See: https://www.bloomberg.com/news/features/2019-07-18/amazon-s-most-ambitious-research-project-is-a-convenience-store.

105 New York Times *describes passing through the store's turnstiles:* Wingfield, "Inside Amazon Go."

105 *McKinsey estimates automated checkout will save retailers:* "The Internet of Things: Mapping the Value Beyond the Hype," McKinsey & Company, June 2015. See: https://www.mckinsey.com/~/media/McKinsey/Business%20Functions/McKinsey%20Digital/Our%20Insights/The%20Internet%20of%20Things%20The%20value%20of%20digitizing%20the%20physical%20world/The-Internet-of-things-Mapping-the-value-beyond-the-hype.ashx.

105 *the San Francisco startup v7labs:* "This AI Startup Wants to Automate Every Store Like Amazon Go," *Fast Company*, November 9, 2017. See: https://www.fastcompany.com/40493622/this-ai-startup-wants-to-automate-every-store-like-amazon-go. See also: https://www.v7labs.com/retail.

105 *Alibaba's cashier-less Hema stores were tested in China:* Rebecca Fannin, "Alibaba Beats Amazon to New All-Digital Retail Trend," *Forbes*, September 21, 2018. See: https://www.forbes.com/sites/rebeccafannin/2018/09/21/alibaba-beats-amazon-to-new-all-digital-retail-trend/#6b7660436653.

105 *Smart shelf technology:* "The Future of Retail: Shopping and the Smart Shelf," *Intel.* See: https://www.intel.com/content/www/us/en/retail/digital-retail-futurecasting-report.html.

106 *Back in 2015, a Cisco study found that IoT solutions:* Andrew Meola, "How IoT Logistics Will Revolutionize Supply Chain Management," *Business Insider*, December 21, 2016. See: https://www.businessinsider.com/internet-of-things-logistics-supply-chain-management-2016-10.

The Robots Are Coming, The Robots Are Coming

106 *The Robots Are Coming, The Robots Are Coming:* Daniel Faggella, "Artificial Intelligence In Retail—10 Present and Future Use Cases," Emerj.com, March 28, 2019. See: https://emerj.com/ai-sector-overviews/artificial-intelligence-retail/.

106 *Domino's Robotic Unit:* Introducing DOM (video). See: https://www.youtube.com/watch?v=rb0nxQyv7RU&feature=youtu.be.

106 *Already, it's been rolled out in ten countries:* Mariella Moon, "Domino's Delivery Robots Are Invading Europe," *Engadget,* March 30, 2017.

106 *A dozen or so different delivery bots are currently entering the market. Starship Technologies:* Kayla Mathews, "5 Ways Retail Robots Are Disrupting the

Industry," *Robotics Business Review*, August 2, 2018. See: https://www.roboticsbusinessreview.com/retail-hospitality/retail-robots-disrupt-industry/.

106 *Starship has carried out fifty thousand deliveries in over one hundred cities in twenty countries:* Luke Dormehl, "The Rise and Reign of Starship, the World's First Robotic Delivery Provider," *Digital Trends*, May 22, 2019. See: https://www.digitaltrends.com/cool-tech/how-starship-technologies-created-delivery-robots/.

107 *Along similar lines, Nuro, the company cofounded by Jiajun Zhu:* Mark Harris, "Softbank's $940 Million Smaller Robots Could Leap from Driverless Vehicles to Complete Last-Yard Deliveries," *TechCrunch*, March 23, 2019. See: https://techcrunch.com/2019/03/23/how-nuro-plans-to-spend-softbanks-money/.

107 *which it's been doing for select Kroger stores since 2018:* Kyle Wiggers, "Nuro Expands Kroger Driverless Deliveries to Houston," *Venture Beat*, March 14, 2019. See: https://venturebeat.com/2019/03/14/nuro-expands-driverless-delivery-partnership-with-kroger-to-houston/.

107 *in 2019, Domino's also partnered with Nuro:* Alex Davies, "Nuro's Pizza Robot Will Bring You a Domino's Pie," *Wired*, June 17, 2019. See: https://www.wired.com/story/nuro-dominos-pizza-delivery-self-driving-robot-houston/.

107 *Prime Air:* "First Prime Air Delivery." See: https://www.amazon.com/Amazon-Prime-Air/b?ie=UTF8&node=8037720011.

107 *7-Eleven:* Nicole Lee, "7-Eleven Has Already Made 77 Deliveries by Drone," *Engadget*, December 20, 2016. See: https://www.engadget.com/2016/12/20/7-eleven-has-already-made-77-deliveries-by-drone/?guccounter=1.

107 *Walmart:* Mary Hanbury, "There's a Way Walmart Could Beat Amazon When It Comes to Speedy Delivery, and New Data Shows It's Going All In," *Business Insider*, June 18, 2019. See: https://www.businessinsider.com/walmart-invests-in-drones-as-amazon-delivery-war-heats-up-2019-6.

107 *Google:* Mihir Zaveri, "Wing, Owned by Google's Parent Company, Gets First Approval for Drone Deliveries in U.S.," *New York Times*, April 23, 2019. See: https://www.nytimes.com/2019/04/23/technology/drone-deliveries-google-wing.html. See also: "Transforming the Way Goods Are Transported," https://x.company/projects/wing/.

107 *Alibaba:* "China Is on the Fast Track to Drone Deliveries," *Bloomberg*, July 3, 2018. See: https://www.bloomberg.com/news/features/2018-07-03/china-s-on-the-fast-track-to-making-uav-drone-deliveries.

107 *head of the FAA's drone-integration department:* "Drone Deliveries Really Are Coming Soon, Officials Say," *Drive*, March 14, 2018. See: https://www.thedrive.com/tech/19239/drone-deliveries-really-are-coming-soon-officials-say.

107 *SoftBank introduced Pepper:* Erico Guizzo, "How Aldebaran Robotics Built Its Friendly Humanoid Robot, Pepper," *IEEE Spectrum*, December 26, 2014. See: https://spectrum.ieee.org/robotics/home-robots/how-aldebaran-robotics-built-its-friendly-humanoid-robot-pepper.

107 *Over twelve thousand Peppers:* Bill Streeter, "Seriously Successful Results from HSBC Bank's Branch Robot Rollout," *Financial Brand*, June 5, 2019, See: https://thefinancialbrand.com/84245/hsbc-banks-branch-robot-pepper-digital-transformation-phygital/. See also: Parmy Olson, "Softbank's Robotics Business Prepares to Scale Up," *Forbes*, May 30, 2018, https://www.forbes.com/sites/parmyolson/2018/05/30/softbank-robotics-business-pepper-boston-dynamics/#2579283e4b7f.

107 *Walmart uses shelf-stocking robots:* Sarah Nassauer, "Walmart Is Rolling Out the Robots," *Wall Street Journal*, April 9, 2019. See: https://www.wsj.com/articles/walmart-is-rolling-out-the-robots-11554782460.

107 *Best Buy uses a robo-cashier:* Kavita Kumar, "Best Buy Tests Robot at New York Store," *Star Tribune*, September 26, 2015. See: http://www.startribune.com/best-buy-tests-robot-at-new-york-store/329583301/.

107 *LoweBot:* "LoweBot." See: http://www.lowesinnovationlabs.com/lowebot.

108 *out $775 million for Kiva Systems:* Evelyn M. Rusli, "Amazon.com to Acquire Manufacturer of Robotics," *New York Times*, March 19, 2012. See: https://dealbook.nytimes.com/2012/03/19/amazon-com-buys-kiva-systems-for-775-million/?mtrref=undefined&gwh=A926616EBBF3A219E03216397142BB8B&gwt=pay&assetType=REGIWALL.

108 *Kiva robots:* Sam Shead, "Amazon Now Has 45,000 Robots in Its Warehouses," *Business Insider*, January 3, 2017. See: https://www.businessinsider.com/amazons-robot-army-has-grown-by-50-2017-1.

108 *306 items* per second*:* Jay Yarow, "Amazon Was Selling 306 Items Every Second at Its Peak This Year," *Business Insider*, December 27, 2012. https://www.businessinsider.com/amazon-holiday-facts-2012-12.

108 *Order jeans from the Gap:* Bob Trebilcock, "Resilience and Innovation at Gap Inc.," *Modern Materials Handling*, November 12, 2018. See: https://www.mmh.com/article/resilience_and_innovation_at_gap_inc.

108 *Kindred robot:* "Kindred Robot." See: https://www.kindred.ai/. (Author note: Peter's VC firm is an investor.)

108 *the House of Representatives already passed the bill:* Nandita Bose, "House Passes Bill to Raise Federal Minimum Wage to $15 An Hour," Reuters, July 18, 2019. See: https://www.reuters.com/article/us-usa-congress-minimum-wage/house-passes-bill-to-raise-federal-minimum-wage-to-15-an-hour-idUSKCN1UD2DV.

3-D Printing and Retail

109 *Ministry of Supply:* "Ministry of Supply." See: https://ministryofsupply.com/. See also: Richard Kestenbaum, "3-D Printing In-Store Is Very Close and Retailers Need to Address It," *Forbes*, April 6, 2017, https://www.forbes.com/sites/richardkestenbaum/2017/04/06/3-D-printing-in-store-is-very-close-and-retailers-need-to-address-it/#4ba78ea333b4.

109 *As* TechCrunch *summarized:* Rip Empson, "With Tech from Space, Ministry of Supply Is Building the Next Generation of Dress Shirts," *TechCrunch*, July 1, 2012. See: https://techcrunch.com/2012/06/30/ministry-of-supply/.

109 *Danit Peleg's:* "Danit Peleg." See: https://danitpeleg.com/.

109 *Reebok:* Brian Lord, "Reebok's 3-D Printed Shoe Line Dashes into Production," *3-D Printing Industry*. August 3, 2018, See: https://3-Dprintingindustry.com/news/reeboks-3-D-printed-shoe-line-dashes-into-production-137497/.

109 *New Balance:* Tyler Koslow, "New Balance Announces 3-D Printed Midsoles in New Running Shoe Line," *3-D Printing Industry*, November 19, 2015. See: https://3-Dprintingindustry.com/news/new-balance-announces-3-D-printed-midsoles-in-new-running-shoe-line-62313/.

110 *Staples, the office supply company:* Scott J. Grunewald, "Staples' New Sculpteo-Powered Online 3-D Printing Service Launches," *3-D Print.com*, September 17, 2015. See: https://3-Dprint.com/96380/staples-sculpteo-launched/.

110 *Leroy Merlin:* Beau Jackson, "Interview: How ZMorph 3-D Printers Are Helping the Leroy Merlin Bricolab Movement in Brazil," *3-D Printing Industry*, January 28, 2019. See: https://3-Dprintingindustry.com/news/interview-how-zmorph-3-D-printers-are-helping-the-leroy-merlin-bricolab-movement-in-brazil-147975/.

Retail's Last Hope

111 *"Welcome to the Experience Economy":* B. Joseph Pine II, "Welcome to the Experience Economy," *Harvard Business Review*, July-August 1998. See: https://hbr.org/1998/07/welcome-to-the-experience-economy.

112 *group's ten-year vision for the future of retail:* Christophe Cuvillier, "Destination 2028," *The URW 2018 Report*. See: https://report.urw.com/2018/.

112 *there are over eleven hundred malls and forty thousand shopping centers:* "Shopping Center," Encyclopedia.com. See: https://www.encyclopedia.com/reference/encyclopedias-almanacs-transcripts-and-maps/shopping-center.

112 *spanning 2.5 million square feet and housing five hundred stores:* "Mall of America by the Numbers." See: https://www.mallofamerica.com/upload/FactSheets_2016.pdf.

112 *China's biggest mall:* "The Largest Shopping Malls in Asia," World Atlas. See: https://www.worldatlas.com/articles/the-largest-shopping-malls-in-asia.html.

No More Shopping Malls

114 *Levi's:* Pamela Kokoszka, "Is 3-D Body Scanning the Future Of Fashion?": Verdict. See: https://www.verdict.co.uk/3-D-body-scanning-fashion-future/.

114 *Bloomingdales:* Marc Bain, "Could 3-D Body Scanning Mean Never Entering Another Dressing Room Again," *Quartz,* September 9, 2015. See: https://qz.com/497259/could-3-D-body-scanning-mean-never-entering-another-dressing-room-again/.

114 *Boss:* Lydia Mageean, "3-D Technology: A New Dimension for Fashion," whichPLM, November 22, 2018. See: https://www.whichplm.com/3-D-technology-a-new-dimension-for-fashion/.

114 *Armani:* "Bodi.Me." See: http://bodi.me/.

114 *Bombfell:* Ryan Lawler, "500 Startups-Backed Bombfell Helps Nerds Get Stylish, for Just $69 a Month," *TechCrunch,* June 14, 2012. See: https://techcrunch.com/2012/06/14/bombfell/.

114 *Amazon acquiring the 3-D body-scanning startup Body Labs:* Natasha Lomas, "Amazon Has Acquired 3-D Body Model Startup, Body Labs, for $50M-$70M," *TechCrunch,* October 3, 2017. See: https://techcrunch.com/2017/10/03/amazon-has-acquired-3-D-body-model-startup-body-labs-for-50m-70m/.

114 *Alibaba's FashionAI:* Christine Chou, "New Alibaba Concept Store Teases Future of Fashion Retail," July 4, 2018. See: https://www.alizila.com/new-alibaba-concept-store-teases-future-of-fashion-retail/.

114 *Amazon's shopping algorithm:* "Real-Time Product Recommendations," Amazon. See: https://aws.amazon.com/mp/scenarios/bi/recommendation/.

114 *Hololux:* Matt Brown, "How Microsoft Is Shaking Up Fashion with Mixed Reality, AI, and IoT," Windows Central, June 9, 2018. See: https://www.windowscentral.com/microsoft-fashion-mixed-reality-ai-iot.

Chapter Six: The Future of Advertising

Madder Men

117 *the disruption it would bring to advertising:* Ryan Avent, *The Wealth of Humans: Work, Power, and Status in the Twenty-First Century* (St. Martins, 2016), p. 3.

117 *In 2017, Google's ad campaign revenue:* See: https://www.statista.com/statistics/266249/advertising-revenue-of-google/.

118 *Facebook's reached over $39 billion:* See: https://www.statista.com/statistics/271258/facebooks-advertising-revenue-worldwide/.

118 *In 2018, the global advertising industry surpassed $550:* See: https://www.statista.com/statistics/236943/global-advertising-spending/.

118 *Google's valuation:* Matthew Lynley, "Google Joins the Race to $1 Trillion," *TechCrunch,* July 23, 2018. See: https://techcrunch.com/2018/07/23/google-joins-the-race-to-1-trillion/.

118 *Facebook's above $500 billion:* Matt Egan, "Facebook and Amazon Hit $500 Billion Milestone," CNN, July 27, 2017. See: https://money.cnn.com/2017/07/27/investing/facebook-amazon-500-billion-bezos-zuckerberg/index.html.

The Spatial Web

119 *A partnership between Snapchat and Amazon:* Josh Constine, "Snapchat Lets You Take a Photo of an Object to Buy It on Amazon," *TechCrunch,* September 24, 2018. See: https://techcrunch.com/2018/09/24/snapchat-amazon-visual-search/.

119 *Pinterest, meanwhile, has a multitude of visual search tools:* See Pinterest's original announcement: https://newsroom.pinterest.com/en/post/introducing-the-next-wave-of-visual-search-and-shopping.

120 *Google Lens:* See: https://lens.google.com.

120 *IKEA:* Arielle Pardes, "Ikea's New App Flaunts What You'll Love Most About AR," *Wired,* September 20, 2017. See: https://www.wired.com/story/ikea-place-ar-kit-augmented-reality/.

120 *visual searches above a billion queries a month:* Yoram Wurmser, "Visual Search 2018: New Tools from Pinterest, eBay, Google and Amazon Increase Accuracy, Utility," *eMarketer,* September 26, 2018. See: https://www.emarketer.com/content/visual-search-2018.

Sure That Sounds like Mom, But Can You Prove It?

121 *Montreal-based startup Lyrebird:* Partner Content, "How Lyrebird Uses AI to Find Its (Artificial) Voice," *Wired.* See: https://www.wired.com/brandlab/2018/10/lyrebird-uses-ai-find-artificial-voice/. See also: https://lyrebird.ai/.

121 *Thirty sentences is all it takes:* Greg Allen, "AI Will Make Forging Anything Entirely Too Easy," *Wired,* July 1, 2017. See: https://www.wired.com/story/ai-will-make-forging-anything-entirely-too-easy/.

122 *Leo Zou, a member of Baidu's communications team:* Luke Dormehl, "Baidu's New A.I. Can Mimic Your Voice After Listening to It for Just One Minute," *Digital Trends,* February 28, 2018. See: https://www.digitaltrends.com/cool-tech/baidu-ai-emulate-your-voice/.

Deeper Fakes

122 *2018, a YouTube video:* David Mack, "This PSA About Fake News from Barack Obama Is Not What It Appears," *BuzzFeed News,* April 17, 2018. See: https://www.buzzfeednews.com/article/davidmack/obama-fake-news-jordan-peele-psa-video-buzzfeed.

122 *the dangers of deepfakes:* Technology, "The Real Danger of DeepFake Videos Is That We May Question Everything," *NewScientist,* August 29, 2018. See: https://www.newscientist.com/article/mg23931933-200-the-real-danger-of-deepfake-videos-is-that-we-may-question-everything/. See also: Oscar Swartz, "You Thought Fake News Was Bad? Deep Fakes Are Where Truth Goes to Die," *Guardian,* November 12, 2018, https://www.theguardian.com/technology/2018/nov/12/deep-fakes-fake-news-truth.

122 *Carnegie Mellon University:* See Carnegie Mellon's original article: https://www.cmu.edu/news/stories/archives/2018/september/deep-fakes-video-content.html.

Goodbye Advertising, Hello JARVIS

123 *Shopping JARVIS:* Vala Afshar, "How AI-Powered Commerce Will Change Shopping," *ZDNet,* December 21, 2018. See: https://www.zdnet.com/article/how-ai-powered-commerce-will-change-shopping/.

Chapter Seven: The Future of Entertainment

Going Digital

126 *In 1999, the year Netflix launched, they had 239,000 subscribers:* Reed Hastings, "How I Did It: Reed Hastings, Netflix," *Inc. Magazine,* December 1, 2005. See: https://www.inc.com/magazine/20051201/qa-hastings.html.

126 *his 2007 choice:* Miguel Helft, "Netflix to Deliver Movies to the PC," *New York Times,* January 16, 2007. See: https://www.nytimes.com/2007/01/16/technology/16netflix.html.

126 *subscriptions had rocketed to 137 million:* Mansoor Iqbal, "Netflix Revenue and Usage Statistics (2018)," *Business of Apps,* February 27, 2019. See: https://www.businessofapps.com/data/netflix-statistics/.

126 *Fifty-one percent of all streaming subscriptions:* Ibid.

126 *$4.5 billion in annual revenue:* Anthony Ha, "Netflix Added 9.6M Subscribers in Q1, with Revenue of $4.5B," *TechCrunch,* April 16, 2019. See: https://techcrunch.com/2019/04/16/netflix-q1-earnings/.

126 *$150 billion in market cap:* "Netflix Market Cap 2006-2019 | NFLX." Macrotrends.com. See: https://www.macrotrends.net/stocks/charts/NFLX/netflix/market-cap.

126 *$6.2 billion on original:* Arne Alsin, "The Future of Media: Disruptions, Revolutions and the Quest for Distribution," *Forbes*, July 19, 2018. See: https://www.forbes.com/sites/aalsin/2018/07/19/the-future-of-media-disruptions-revolutions-and-the-quest-for-distribution/#5a2ca52d60b9.

126 *Netflix's doubled their spend to $13 billion:* "Netflix Is Moving Television Beyond Time-Slots and National Markets," *Economist*, June 30, 2018. See: https://www.economist.com/briefing/2018/06/30/netflix-is-moving-television-beyond-time-slots-and-national-markets.

126 *Netflix's war chest produced eighty new features and over seven hundred new TV shows:* Todd Spangler, "Netflix Eyeing Total of About 700 Original Series in 2018," *Variety*, February 27, 2018.

126 *Blockbuster passed up the opportunity to buy Netflix:* Marc Graser, "Epic Fail: How Blockbuster Could Have Owned Netflix," *Variety*, November 12, 2013. See: https://variety.com/2013/biz/news/epic-fail-how-blockbuster-could-have-owned-netflix-1200823443/.

127 *Apple spent over $1 billion on original programming:* Tripp Mickle, "Apple Readies $1 Billion War Chest for Hollywood Programming," *Wall Street Journal*, August 16, 2017. See: https://www.wsj.com/articles/apple-readies-1-billion-war-chest-for-hollywood-programming-1502874004.

127 *Amazon dropped $5 billion:* Adam Levy, "Here's Exactly How Much Amazon Is Spending on Video and Music Content," *Motley Fool*, April 30, 2019. See: https://www.fool.com/investing/2019/04/30/heres-how-much-amazon-is-spending-on-video-music.aspx.

127 *TV ad sales and box office revenue produce just shy of $300 billion a year:* For global TV revenues see: "Global TV Revenues Grow to $265 Billion," *Broadband TV News*, October 22, 2018, https://www.broadbandtvnews.com/2018/10/22/global-tv-revenues-grow-to-265-billion/. For global box office revenues see: Pamela McClintock, "Global Box Office Revenue Hits Record $41B in 2018, Fueled by Diverse U.S. Audiences," *Hollywood Reporter*, March 21, 2019, https://www.hollywoodreporter.com/news/global-box-office-revenue-hits-record-41b-2018-fueled-by-diverse-us-audiences-1196010.

The Rise of the Uber-Creator

128 *Jawed Karim posted "Me at the Zoo,":* "Me at the Zoo" (video), April 23, 2005. See: https://www.youtube.com/watch?v=jNQXAC9IVRw.

128 *Brazilian soccer phenom Ronaldinho:* "Ronaldinho Nike Ad" (video), August 2, 2006. See: https://www.youtube.com/watch?v=i_JS1YG8H2c.

128 *$3.5 million investment from Sequoia Capital:* Miguel Helft, "Venture Firm Shares a YouTube Jackpot," *New York Times*, October 10, 2006. See: https://www.nytimes.com/2006/10/10/technology/10payday.html.

128 *$1.65 billion to purchase YouTube:* Ibid.

128 *billions of people watch billions of videos on the site:* "Over One Billion Users." See: https://www.youtube.com/about/press/.

128 *shows like* Binging with Babish*:* "Binging with Babish" (video). See: https://www.youtube.com/user/bgfilms.

128 *Or* Cooking with Dog*:* "Cooking with Dog" (video). See: https://www.youtube.com/user/cookingwithdog.

128 *in to* My Drunk Kitchen: "My Drunk Kitchen" (video). See: https://www.youtube.com/user/MyHarto.

129 *And these new stars are bringing in big bucks:* Natalie Robehmed, "Highest-Paid YouTube Stars 2018: Markiplier, Jake Paul, Pewdiepie and More," *Forbes*, December 3, 2018. See: https://www.forbes.com/sites/natalierobehmed/2018/12/03/highest-paid-youtube-stars-2018-markiplier-jake-paul-pewdiepie-and-more/#22c2828e909a.

129 *Bambuser:* "Bambuser." See: https://bambuser.com.

130 *short film* Sunspring*:* "Sunspring" (video). See: https://www.youtube.com/watch?v=LY7x2Ihqjmc.

130 *upcoming thriller* Morgan*:* "BM Creates First Movie Trailer by AI [HD] | 20th Century FOX," August 31, 2016. See: https://www.youtube.com/watch?v=gJEzuYynaiw.

130 *an AI that creates Choose Your Own Adventure–style stories for video games:* Matthew Guzdial, "Crowdsourcing Open Interactive Narrative." See: https://www.cc.gatech.edu/~riedl/pubs/guzdial-fdg15.pdf. See also: "Artificial Intelligence System for Crowdsourcing Interactive Fiction" (video), September 1, 2016. See: https://www.youtube.com/watch?time_continue=1&v=znqw17aOrCs.

From Passive to Active

131 *Video games with user-generated gameplay content:* "Category: Video Games with User-Generated Gameplay Content," Wikipedia. See: https://en.wikipedia.org/wiki/Category:Video_games_with_user-generated_gameplay_content.

131 *MashUp Machine:* "Mashup Machine." See: http://mashupmachine.io/.

131 *researchers at the University of California, Berkeley:* Caroline Chan, "Everybody Dance Now," Arvix.org, August 22, 2018. See: https://arxiv.org/pdf/1808.07371v1.pdf.

132 *Robbins teamed up:* Tony Robbins, author interview, 2018.

132 *Lifekind:* "AI Personas Based on Real People." See: http://www.rivaltheory.com/.

The Holodeck Is Here

133 *Jules Urbach went to high school with Rod Roddenberry:* Jules Urbach, author interview, 2018. See also: Fotis Georgiadis, "The Future Is Now: 'Now We Can Effortlessly Interact with Digital Holographic Objects That Naturally Blend into Everyday Life' with OTOY CEO Jules Urbach & Fotis Georgiadis," *Thrive Global*, January 16, 2019. See: https://thriveglobal.com/stories/the-future-is-now-now-we-can-effortlessly-interact-with-digital-holographic-objects-that-naturally-blend-into-everyday-life-with-otoy-ceo-jules-urbach-fotis-georgiadis/.

133 *Otoy:* "Otoy." See: https://home.otoy.com/.

133 *Light Field Lab:* "Light Field Lab." See: https://www.lightfieldlab.com/. See also: https://variety.com/2018/digital/features/light-field-lab-holographic-display-demo-1203026693/.

134 *High Fidelity:* "High Fidelity Raises $35m to Bring Virtual Reality to 1 Billion People," *High Fidelity*, June 28, 2018. See: https://www.prnewswire.com/news-releases/high-fidelity-raises-35m-to-bring-virtual-reality-to-1-billion-people-300673807.html.

134 *NeoSensory:* "NeoSensory." See: https://neosensory.com/?v=7516fd43adaa. See also: David Eagleman, "Can We Create New Senses for Humans?," TED Talk, https://www.ted.com/talks/david_eagleman_can_we_create_new_senses_for_humans.

135 *Dreamscape:* "Dreamscape." See: https://dreamscapeimmersive.com/. See also: Bryan Bishop, "Dreamscape Immersive Wants to Bring Location-Based VR to the Masses, Starting with a Shopping Mall," *Verge*, January 15, 2019. See: https://www.theverge.com/2019/1/15/18156854/dreamscape-immersive-virtual-reality-los-angeles-walter-parkes-bruce-vaughn. (Author note: Peter's VC firm is an investor.)

This Time It's Personal

136 *affective computing:* Geovany A. Ramirez, "Color Analysis of Facial Skin: Detection of Emotional State.", University of Texas, See: http://www.cs.utep.edu/ofuentes/papers/emotionSkin_final.pdf.

137 *Affectiva:* "Affectiva." See: https://www.affectiva.com/. See also Samar Marwan, "Rana El Kaliouby CEO of Affectiva Is Training Robots to Read Feelings," *Forbes*, November 29, 2018, https://www.forbes.com/sites/samarmarwan/2018/11/29/affectiva-emotion-ai-ceo-rana-el-kaliouby/#7d5f8c5c1572.

137 *thriller game* Nevermind: "Finding the Essence of Fear in Nevermind," July 21, 2016. See: https://blog.affectiva.com/finding-the-essence-of-fear-in-nevermind.

137 *Lightwave:* Tom Foster, "Ready or Not, Companies Will Soon Be Tracking Your Emotions," *Inc. Magazine*, July 2016. See: https://www.inc.com/magazine/201607/tom-foster/lightwave-monitor-customer-emotions.html.

137 *Ubimo:* "Ubumo." See: https://www.ubimo.com/.

137 *Cluep:* "Cluep." See: https://cluep.com/.

137 *they were turning the 1980s Choose Your Own Adventure:* Ricardo Lopez, " 'Choose Your Own Adventure' Interactive Movie in the Works at Fox," *Variety*, April 26, 2018. See: https://variety.com/2018/film/news/choose-your-own-adventure-interactive-1202788741/.

138 Hollywood Reporter *called it:* Josh Spiegel, "Does a 'Choose Your Own Adventure' Movie Sound Appealing?," *Hollywood Reporter*. See: https://www.hollywoodreporter.com/heat-vision/choose-your-own-adventure-movie-does-it-sound-appealing-1106999.

Here, There, and Everywhere

139 *Magic Leap:* Rony Abovitz, author interview, 2018. See also: "Magic Leap," https://www.magicleap.com/, and Peter Yang, "The Untold Story of Magic Leap, the World's Most Secretive Startup," *Wired*, https://www.wired.com/2016/04/magic-leap-vr/.

139 *OLEDs (organic light emitting diodes):* "OLED Lighting Products: Capabilities, Challenges, Potential," U.S. Department of Energy, May 2016. See: https://www.energy.gov/sites/prod/files/2016/08/f33/ssl_oled-products_2016.pdf.

139 *Chinese researchers to develop a smartphone:* "Bendy Smartphone Made of Graphene Displayed at China Tech Fair" (video) April 27, 2016. See: https://www.youtube.com/watch?v=ZwuQXfHXsa4. See also: Jon Porter, "Nubia's New Wearable Puts a 4-Inch Flexible Smartphone on Your Wrist," *Verge*, https://www.theverge.com/circuitbreaker/2019/2/25/18240370/nubia-alpha-release-date-price-features-wearable-smartwatch-flexible-display-mwc-2019.

140 *AR is projected to create a $90 billion market:* "For AR/VR 2.0 to Live, AR/VR 1.0 Must Die," Digi-Capital, January 15, 2019. See: https://www.digi-capital.com/news/2019/01/for-ar-vr-2-0-to-live-ar-vr-1-0-must-die/.

140 *I regard [AR] as a big idea like the smartphone:* David Phelan, "Apple CEO Tim Cook: As Brexit Hangs over UK, 'Times Are Not Really Awful, There's Some Great Things Happening'," *Independent*. See: https://www.independent.co.uk/life-style/gadgets-and-tech/features/apple-tim-cook-boss-brexit-uk-theresa-may-number-10-interview-ustwo-a7574086.html.

140 *Nintendo released Pokémon GO:* Lauren Munsi, "Pokémon GO Surpasses the 1 Billion Downloads Milestone," *Nintendo Wire*, July 31, 2019. See: https://nintendowire.com/news/2019/07/31/pokemon-go-surpasses-the-1-billion-downloads-milestone/.

140 *With 5 million daily users, 65 million monthly users, and over $2 billion in revenue:* Mansoor Iqbal, "Pokémon GO Revenue and Usage Statistics (2019)," *Business of Apps*, May 10, 2019. See: https://www.businessofapps.com/data/pokemon-go-statistics/.

140 *Mojo Vision:* See: https://mojo.vision/. See also: Dean Takahashi, "Mojo Vision Reveals the World's Smallest and Densest Micro Display," *Venture Beat*, https://venturebeat.com/2019/05/30/mojo-vision-reveals-the-worlds-smallest-and-densest-micro-display/. (Author note: Peter's VC firm is an investor.)

141 *we've seen EEG-based BCI devices start to invade gaming:* Ahn Minkyu, "A Review of Brain-Computer Interface Games and an Opinion Survey from Researchers, Developers and Users," *Sensors*, August 2014, pp. 14601–14633, doi: 10.3390/s140814601. See: https://www.ncbi.nlm.nih.gov/pmc/articles/PMC4178978/.

141 *BrainNet:* Jiang Preston Linxing, "BrainNet: A Multi-Person Brain-to-Brain Interface for Direct Collaboration Between Brains," *Human-Computer Interaction*, May 22, 2019. See: https://arxiv.org/abs/1809.08632.

141 *short film entitled* The Moment*:* Richard Ramchurn, "Now Playing: A Movie You Control with Your Mind," *MIT Technology Review*, May 25, 2018. See: https://www.technologyreview.com/s/611189/now-playing-a-movie-you-control-with-your-mind/.

Chapter Eight: The Future of Education

The Quest for Quantity and Quality

143 *1.6 million teachers:* Read the full U.S. Department of Education report, "Our Future, Our Teachers: The Obama Administration's Plan for Teacher Education Reform and Improvement," here: https://www.ed.gov/sites/default/files/our-future-our-teachers.pdf.

143 *UNESCO estimates:* UNESCO, "The World Needs Almost 69 Million New Teachers to Reach the 2030 Education Goals," *UNESCO Institute for Statistics*, no. 39 (October 2016). See: http://uis.unesco.org/sites/default/files/documents/fs39-the-world-needs-almost-69-million-new-teachers-to-reach-the-2030-education-goals-2016-en.pdf.

143 *Our modern educational system:* For a great overview of these issues, see Ken Robinson, *Out of Our Minds* (Capstone, 2011).

144 *2015 study by the US Department of Education:* Read former U.S. Deputy Secretary Tony Miller's full speech from July 2011 here: https://www.ed.gov/news/speeches/partnering-education-reform.

144 *half of those dropouts citing* boredom: Read the full "The Silent Epidemic Perspectives of High School Dropouts" report here: https://docs.gatesfoundation.org/documents/thesilentepidemic3-06final.pdf.

One Billion Android Teachers Per Year

144 *In 2012, Nicholas Negroponte:* David Talbot, "Given Tablets but No Teachers, Ethiopian Children Teach Themselves," *MIT Technology Review*, October 29, 2012. See: https://www.technologyreview.com/s/506466/given-tablets-but-no-teachers-ethiopian-children-teach-themselves/.

145 *One Laptop per Child:* See: http://one.laptop.org/.

145 *Negroponte told the* MIT Review*:* Ibid.

145 *Sugata Mitra:* Watch Sugata Mitra's full "Kids Can Teach Themselves" TED Talk here: https://www.ted.com/talks/sugata_mitra_shows_how_kids_teach_themselves?language=en.

146 *Ed McNierney, the nonprofit's CTO, told the* MIT Review*:* Ibid.

146 *Global Learning XPRIZE:* See: https://www.xprize.org/prizes/global-learning.

147 *over a billion Android handsets manufactured each year:* According to Gartner, over 1.5 billion smartphones were sold in 2017. Of these, 1.32 billion used the Android operating system. From: Gartner Newsroom, Press Releases, "Gartner Says Worldwide Sales of Smartphones Recorded First Ever Decline During the Fourth Quarter of 2017," https://www.gartner.com/en/newsroom/press-releases/2018-02-22-gartner-says-worldwide-sales-of-smartphones-recorded-first-ever-decline-during-the-fourth-quarter-of-2017.

The Ultimate Field Trip

147 *Philip Rosedale and his team at High Fidelity:* Philip Rosedale, author interview, 2018. Check out this video of the tour: https://www.youtube.com/watch?v=zb2NUs0IDm4ut.

148 *multi-sensory learning:* Bailey, F. & Pransky, K. (2015, July 16). Implications and applications of the latest brain research for learners and teachers [Webinar]. In *Association for Supervision and Curriculum Development Webinar Series.* Retrieved from: http://www.ascd.org/professional-development/webinars/implications-and-applications-of-brain-research- webinar.aspx.

148 *Jeremy Bailenson:* See his profile here: https://vhil.stanford.edu/faculty-and-staff/.

148 *VR can quickly and significantly shift our attitudes and actions:* Fernanda Herrera, "Building Long-Term Empathy: A Large-Scale Comparison of

Traditional and Virtual Reality Perspective-Taking," *PLoS ONE* 13, no. 10 (October 17, 2018). See: https://vhil.stanford.edu/mm/2018/11/herrera-pone-building-long-term-empathy.pdf.

148 *USC, psychologist Skip Rizzo:* See his profile here: http://ict.usc.edu/profile/albert-skip-rizzo/.

149 *full range of anxiety disorders:* Jessica Maples-Keller, "The Use of Virtual Reality Technology in the Treatment of Anxiety and Other Psychiatric Disorders," *Harvard Review of Psychiatry* 25, no. 3 (May-June 2017): 103–113. See: https://www.ncbi.nlm.nih.gov/pmc/articles/PMC5421394/.

School 2030

149 *the novel* The Diamond Age*:* Neal Stephenson, *The Diamond Age: Or, a Young Lady's Illustrated Primer* (Spectra, 1995).

149 *chief futurist at Magic Leap:* Davey Alba, "Sci-Fi Author Neal Stephenson Joins Mystery Startup as 'Chief Futurist,'" *Wired*, December 16, 2014. See: https://www.wired.com/2014/12/neal-stephenson-magic-leap/.

Chapter Nine: The Future of Healthcare

Martine and the Moonshots

151 *Martine Rothblatt:* Martine Rothblatt, author interview, 2018. For more background on Martine, see: Neely Tucker, "Martine Rothblatt: She Founded Siriusxm, a Religion and a Biotech. For Starters," *Washington Post*, December 12, 2014, https://www.washingtonpost.com/lifestyle/magazine/martine-rothblatt-she-founded-siriusxm-a-religion-and-a-biotech-for-starters/2014/12/11/5a8a4866-71ab-11e4-ad12-3734c461eab6_story.html, and Tina Reed, "Martine Rothblatt's Theory of Evolution," *Puget Sound Business Journal*, April 3, 2018, https://www.bizjournals.com/seattle/bizwomen/news/latest-news/2018/04/martine-rothblatts-theory-of-evolution.html.

152 *United Therapeutics:* See: https://www.unither.com/.

153 *the number of patients now living with pulmonary hypertension has climbed from two thousand to forty thousand:* Chloe Sorvino, "How CEO Martine Rothblatt Turns Moonshots into Earthshots," *Forbes*, June 20, 2018. See also: Tucker, "Martine Rothblatt."

153 *only two thousand lungs:* According to the United Network for Organ Sharing, in the US, 2,530 lungs were transplanted in 2018. You can find more transplant data here: https://unos.org/data/transplant-trends/.

153 *half-million people die of lung failure:* Rothblatt, Ibid.

153 *80 percent of those donated for transplant end up in trash cans:* Ibid.

153 *"ex vivo lung perfusion":* For an overview of ex vivo lung perfusion technology, see: George Makdisis, "Ex Vivo Lung Perfusion Review of a Revolutionary Technology," *Annals of Translational Medicine* 5, no. 17 (2017), https://www.ncbi.nlm.nih.gov/pmc/articles/PMC5599284/. See also: this press release about FDA approval for one of United Therapeutics' ex vivo lung perfusion treatments: https://www.biospace.com/article/releases/united-therapeutics-announces-fda-approval-of-xps-and-steen-solution-used-to-perform-centralized-ex-vivo-lung-perfusion-services/.

153 *this procedure has saved thousands of lives:* Rothblatt, ibid.

153 *xenotransplantation:* The *New York Times* wrote a detailed piece about xenotransplantation. You can read more here: Tom Clynes, "20 Americans Die Each Day Waiting for Organs. Can Pigs Save Them?,"*New York Times Magazine,* November 11, 2018, https://www.nytimes.com/interactive/2018/11/14/magazine/tech-design-xenotransplantation.html?emc=edit_nn_20181114&nl=morning-briefing&nlid=8215381320181114&te=1&mtrref=undefined&assetType=PAYWALL. See also: this *Nature* article: Sara Reardon, "New Life for Pig Organs," *Nature,* November 12, 2015, https://www.nature.com/news/polopoly_fs/1.18768!/menu/main/topColumns/topLeftColumn/pdf/527152a.pdf?origin=ppub.

153 *teaming up with Craig Venter and Synthetic Genomics:* Rothblatt, ibid.

154 *3-D printing an artificial lung scaffold:* Antonio Regalado, "Inside the Effort to Print Lungs and Breathe Life into Them with Stem Cells," *MIT Technology Review,* June 28, 2018. See: https://www.technologyreview.com/s/611236/inside-the-effort-to-print-lungs-and-breathe-life-into-them-with-stem-cells/.

154 *Beta Technologies':* See: https://www.beta.team/. See also: Eric Adams, "Beta Technologies, a Vermont Air Taxi Start-Up, Might Be About to Change the Aviation World," *Drive,* January 11, 2019, https://www.thedrive.com/tech/25914/beta-technologies-a-vermont-e-vtol-air-taxi-start-up-might-be-about-to-change-the-aviation-world. (Author note: Peter's VC firm is an investor.)

Turning Sick Care into Healthcare

155 *$210 billion per year on procedures patients don't need:* Marshall Allen, "Unecessary Tests and Treatments," *Scientific American*, November 29, 2017, See: https://www.scientificamerican.com/article/unnecessary-tests-and-treatment-explain-why-health-care-costs-so-much/. See also: https://www.nap.edu/read/13444/chapter/1#xvii.

155 *Out of every five thousand new drugs introduced* Fact Sheet*:* New Drug Development, California Biomedical Research Assocaition See: https://studylib.net/doc/8182066/fact-sheet-new-drug-development-process.

155 *takes twelve years to get from lab to patient at a cost of $2.5 billion:* Ibid.

155 *Americans spend an average of $10,739 per person per year on healthcare:* The United States Government's Centers for Medicare & Medicaid Services presents healthcare expenditure data on their website, found here: https://www.cms.gov/research-statistics-data-and-systems/statistics-trends-and-reports/nationalhealthexpenddata/nationalhealthaccountshistorical.html.

155 *by 2027, this single industry will consume nearly 20 percent of the US GDP:* Read the full press release from the United States Government's Centers for Medicare & Medicaid Services here: https://www.cms.gov/newsroom/press-releases/2016-2025-projections-national-health-expenditures-data-released.

155 *"it will be about health":* Lizzy Gurdus, "Tim Cook: Apple's Greatest Contribution Will Be 'About Health,'" CNBC, January 8, 2019. See: https://www.cnbc.com/2019/01/08/tim-cook-teases-new-apple-services-tied-to-health-care.html.

156 *Racing Apple are Google, Amazon, Facebook, Samsung, Baidu, Tencent, and others:* CB Insights does an excellent job summarizing the big-tech healthcare ecosystem in these two reports: https://www.cbinsights.com/research/top-tech-companies-healthcare-investments-acquisitions/ and https://www.cbinsights.com/research/google-amazon-apple-health-insurance/.

DIY Diagnostics

156 *Oura ring:* See: https://ouraring.com. (Author note: Peter's VC firm is an investor.)

157 *Exo Imaging's AI-enabled, cheap, handheld ultrasound 3-D imager:* Exo recently emerged from stealth with a $35 million raise. Read the full press release here: https://www.businesswire.com/news/home/20190805005114/en/Exo-Imaging-Emerges-Stealth-Mode-35M-Series|. See also: https://www.exo-imaging.com/. (Author note: Peter's VC firm is an investor.)

157 *Mary Lou Jepsen's startup, Openwater:* Mary Lou Jepsen gave a TED Talk about Openwater that you can find here: https://www.youtube.com/watch?v=awADEuv5vWY. See also: https://www.openwater.cc/about-us. (Author note: Peter's VC firm is an investor.)

157 *Apple's fourth-generation iWatch:* Read Apple's full press release here: https://www.apple.com/newsroom/2018/09/redesigned-apple-watch-series-4-revolutionizes-communication-fitness-and-health/.

157 *Final Frontier Medical Devices' DxtER:* See: XPrize.com, https://tricorder.xprize.org/prizes/tricorder/articles/family-led-team-takes-top-prize-in-qualcomm-tricor.

158 *a field predicted to become a $102 billion market by 2022:* Prediction by Zion Market Research Group in their report titled "mHealth Market by Devices, by Stakeholder, by Service, by Therapeutics and by Applications: Global

Industry Perspective, Comprehensive Analysis and Forecast, 2014–2022," found here: https://www.zionmarketresearch.com/report/mhealth-market. This press release presents some data from the report: https://www.globenewswire.com/news-release/2017/11/15/1193431/0/en/mHealth-Market-Size-Projected-to-Reach-USD-102-43-Billion-by-2022-Zion-Market-Research.html.

158 *Woebot:* Megan Molteni, "The Chatbot Therapist Will See You Now," *Wired*, June 7, 2017. See: https://www.wired.com/2017/06/facebook-messenger-woebot-chatbot-therapist/. See also: https://woebot.io/.

158 *Human Longevity Inc.:* See: https://www.humanlongevity.com/.

Reading, Writing, and Editing the Code of Life

159 *Jason Vassy, a professor of medicine at Boston's Brigham and Women's Hospital:* Rob Stein, "Routine DNA Sequencing May Be Helpful and Not as Scary as Feared," NPR, June 26, 2017. See: https://www.npr.org/sections/health-shots/2017/06/26/534338576/routine-dna-sequencing-may-be-helpful-and-not-as-scary-as-feared.

159 *results published in the* Annals of Internal Medicine*:* Jason Vasey, "The Impact of Whole-Genome Sequencing on the Primary Care and Outcomes of Healthy Adult Patients: A Pilot Randomized Trial," *Annals of Internal Medicine* 67, no. 3 (2017): 159–169. See: https://annals.org/aim/fullarticle/2633848/impact-whole-genome-sequencing-primary-care-outcomes-healthy-adult-patients.

159 *All of Us project:* Details of the All of Us project and funding can be found on the NIH website here: https://allofus.nih.gov/.

159 *Harvard geneticist George Church recently founded Nebula Genomics:* "Q&A: George Church and Company on Genomic Sequencing, Blockchain, and Better Drugs," *Science,* February 8, 2018. See: https://www.sciencemag.org/news/2018/02/q-george-church-and-company-genomic-sequencing-blockchain-and-better-drugs. For more information on Nebula Genomics, see their website here: https://nebula.org/.

160 *genetically engineered cocaine resistance into mice:* Yuanyuan Li, "Genome-Edited Skin Epidermal Stem Cells Protect Mice from Cocaine-Seeking Behaviour and Cocaine Overdose," *Nature Biomedical Engineering* 3 (2019): 105–113. See: https://www.nature.com/articles/s41551-018-0293-z.

160 *switched off the gene responsible for Duchenne muscular dystrophy in dogs:* Leonela Amoasii, "Gene Editing Restores Dystrophin Expression in a Canine Model of Duchenne Muscular Dystrophy," *Science* 362, no. 6410: 86–91. See: https://science.sciencemag.org/content/362/6410/86.

160 *personalized cancer therapies in humans*: Emily Mullin, "FDA Approves Groundbreaking Gene Therapy for Cancer," *MIT Technology Review*, August

30, 2017. See: https://www.technologyreview.com/s/608771/the-fda-has-approved-the-first-gene-therapy-for-cancer/.

160 *field trials were getting under way in Burkina Faso:* Megan Molteni, "Here's the Plan to End Malaria with Crispr-Edited Mosquitos," *Wired*, September 24, 2018. See: https://www.wired.com/story/heres-the-plan-to-end-malaria-with-crispr-edited-mosquitoes/. See also the original journal article: Nikolai Windbichler, "Targeting the X Chromosome During Spermatogenesis Induces Y Chromosome Transmission Ratio Distortion and Early Dominant Embryo Lethality in Anopheles gambiae," *PLoS Genet* 4, no. 12 (2008). See: https://www.ncbi.nlm.nih.gov/pmc/articles/PMC2585807/.

160 *half of the thirty-two thousand most common genetic disorders:* Peter Reuell, "A Step Forward in DNA Base Editing," *Harvard Gazette*, October 25, 2017. See: https://news.harvard.edu/gazette/story/2017/10/a-step-forward-in-dna-base-editing/.

160 *eighteen thousand diseases from our lives:* Liat Ben-Senior, "[Infographic] 10 Most Common Genetic Diseases," *LabRoots*, May 22, 2018. See: https://www.labroots.com/trending/infographics/8833/10-common-genetic-diseases.

The Future of Surgery

160 *2030s, when NASA is planning to launch the first human exploration mission:* Jeff Foust, "Bridenstine Says NASA Planning for Human Mars Missions in 2030s," *Space News*, July 16, 2019. See: https://spacenews.com/bridenstine-says-nasa-planning-for-human-mars-missions-in-2030s/.

160 *.06 percent per person per year:* Richard Summers, "Emergencies in Space," *The Practice of Emergency Medicine/Concepts*, 2005. See: https://pdfs.semanticscholar.org/a102/d4e61620dd77f93639cf47492f7ca6f8c44f.pdf.

161 *Elon Musk once explained:* Watch Elon Musk's speech at the SS R&D Conference on July 19, 2017 here: https://www.youtube.com/watch?v=BqvBhhTtUm4.

161 *Kim is part of the research team behind STAR:* Alan Brown, "Smooth Operator: Robot Could Transform Soft-Tissue Surgery," *Alliance of Advanced Biomedical Engineering*, August 14, 2017. See: https://aabme.asme.org/posts/smooth-operator-robot-could-transform-soft-tissue-surgery.

161 *are roughly 50 million surgeries undertaken in the US each year:* Margaret J. Hall, "Ambulatory Surgery Data from Hospitals and Ambulatory Surgery Centers: United States, 2010," *Centers for Disease Control National Health Statistics Reports*, no. 102 (February 28, 2017). See: https://www.cdc.gov/nchs/data/nhsr/nhsr102.pdf.

161 *fewer than 5 percent are done robotically:* Christina Frangou, "An Eye on Surgical Robots," *General Surgery News*, July 9, 2018.

162 *a partnership between Alphabet and Johnson & Johnson:* See the initial press

release announcing the partnership: https://www.jnj.com/media-center/press-releases/johnson-johnson-announces-formation-of-verb-surgical-inc-in-collaboration-with-verily. For more information on Verb Surgical, See also: https://www.verbsurgical.com/

162 *Bionaut Labs:* See: https://www.bionautlabs.com/. (Author note: Peter's VC firm is an investor.)

162 *90 percent of all potential treatment candidates:* Alex Zhavoronkov, "Artificial Intelligence for Drug Discovery, Biomarker Development, and Generation of Novel Chemistry," *Molecular Pharmaceutics* 15 (2018): 4311,àí4313.

163 *Da Vinci robot:* See: https://www.davincisurgery.com/da-vinci-systems/about-da-vinci-systems.

163 *reveals people with little training creating exceptionally functional prosthetic limbs:* The US National Institutes of Health has a library of 3-D printable CAD files for functional prosthetics found here: https://3-Dprint.nih.gov/collections/prosthetics.

163 *rounded semiconducting material capable of turning light into patterns:* Sung Hyun Park, "3-D Printed Polymer Photodetectors," *Advanced Materials* 30, no. 40 (October 4, 2018). See: https://onlinelibrary.wiley.com/doi/abs/10.1002/adma.201803980.

Cellular Medicine

163 *discovery of stem cells:* For more extensive background on stem cells, see: https://stemcells.nih.gov/info/basics/4.htm.

163 *Robert Hariri:* Robert Hariri, author interview, 2018.

163 *year 2000 discovery that the human placenta houses an abundant supply of stem cells:* Ibid.

164 *CAR-T (for chimeric antigen receptor T-cells) therapy:* The US National Cancer Institute provides a good explanation of CAR-T technology here: https://www.cancer.gov/about-cancer/treatment/research/car-t-cells.

164 *price was roughly half-a-million dollars per patient:* Johnathan Rockoff, "The Million-Dollar Cancer Treatment: Who Will Pay?" *Wall Street Journal,* April 26, 2018. See: https://www.wsj.com/articles/the-million-dollar-cancer-treatment-no-one-knows-how-to-pay-for-1524740401.

164 *Celularity:* See: https://www.celularity.com/.

The Future of Drugs

165 *if a pharmaceutical company wants to create a new medicine:* See the US Food and Drug Administration's overview of the Drug Development Process here: https://www.fda.gov/patients/drug-development-process/step-1-discovery-and-development.

165 *between $2.5 billion and $12 billion to get there:* Joseph DiMasi, "Innovation in the Pharmaceutical Industry: New Estimates of R&D Costs," *Journal of Health Economics* 47 (May 2016): 20–33, https://doi.org/10.1016/j.jhealeco.2016.01.012.

165 *Alex Zhavoronkov:* Alex Zhavoronkov, author interview, 2018. For more information on Insilico Medicine, see: https://insilico.com/. (Author note: Peter's VC firm is an investor.)

165 *AI to significantly speed up the drug discovery process:* Ibid. For more detailed information, see the American Chemical Society's *Molecular Pharmaceutics* special issue on using AI for drug discovery.

166 *Insilico was generating novel molecules in fewer than forty-six days:* Alex Zhavoronkov, "Deep Learning Enables Rapid Identification of Potent DDR1 Kinase Inhibitors," *Nature Biotechnology* 37 (2019): 1038–1040. See: https://www.nature.com/articles/s41587-019-0224-x. (Author note: Peter's VC firm is an investor.)

167 *researchers only managed to find about five new drug targets a year:* Reinhard Renneberg, *Biotechnology for Beginners* (*Academic Press*, 2016), p. 281.

167 *a biannual competition was created:* See: http://predictioncenter.org/.

167 *AlphaFold:* Read the DeepMind blog about AlphaFold here: https://deepmind.com/blog/alphafold/.

Chapter Ten: The Future of Longevity

The Nine Horsemen of Our Apocalypse

169 *Francis Collins:* Dr. Francis Collins shared the stage with Peterat at a 2018 event hosted by the Cura Foundation. Watch the conversation on *Longevity and the Morality of Extreme Life Extension* here: https://www.youtube.com/watch?v=z6i0yTA4zRM. Peter wrote about his experience with Dr. Collins in his blog here: *Peter Diamandis,* "The Morality of Immortality," 2018, https://www.diamandis.com/blog/the-morality-of-immortality.

170 *the nine horsemen of an internal apocalypse:* Carlos Lopez-Otin, "The Hallmarks of Aging," *PMC*, November 23, 2013. See also: "The Future of Aging? The New Drugs & Tech Working to Extend Life & Wellness," *CB Insights*, https://www.ncbi.nlm.nih.gov/pmc/articles/PMC3836174/, and *CB Insights'* report on the future of aging, https://www.cbinsights.com/research/report/future-aging-technology-startups/.

Longevity Escape Velocity

172 *Caenorhabditis elegans, or, as her friends call her,* C. elegans*:* For an overview of *C. elegans*'s role in the study of genetics, see: Claudiu A. Giurumescu,

"Cell Identification and Cell Lineage Analysis," *Methods in Cell Biology*, 2011, doi.org/10.1016/B978-0-12-544172-8.00012-8.

172 C. elegans *was the first organism to have its genes sequenced:* C. elegans Sequencing Consortium, "Genome Sequence of the Nematode C. Elegans: A Platform for Investigating Biology," *Science*, December 11, 1998. See: https://www.ncbi.nlm.nih.gov/pubmed/9851916.

172 *connectome, the wiring diagram of the brain's neurons, mapped:* Steven J. Cook, "Whole-Animal Connectomes of Both *Caenorhabditis elegans* Sexes," *Nature*, July 3, 2019. See: https://www.nature.com/articles/s41586-019-1352-7.

172 *NIH scientists at the Buck Institute for Research on Aging:* See the original paper on this study here: D. Chen, "Germline Signaling Mediates the Synergistically Prolonged Longevity Produced by Double Mutations in daf-2 and rsks-1 in C. elegans," *Cell Reports*, December 26, 2013, doi: 10.1016/j.celrep.2013.11.018. See also: NIH director Dr. Francis Collins's take on the study, "Deciphering Secrets of Longevity, from Worms," *NIH Director's Blog*, January 7, 2014. See the original paper on this study here: https://directorsblog.nih.gov/2014/01/07/deciphering-secrets-of-longevity-from-worms/.

173 *NIH director Francis Collins:* Collins, "Deciphering Secrets."

173 *fifty more genes that seem to trigger age-related decline:* George L. Sutphin, "*Caenorhabditis elegans* Orthologs of Human Genes Differentially Expressed with Age Are Enriched for Determinants of Longevity," *Aging Cell*, April 2017. See: https://onlinelibrary.wiley.com/doi/full/10.1111/acel.12595.

173 *forty-eight years by 1950, then to seventy-two years by 2014:* James Riley and Max Roser, "Life Expectancy by World Region," *Our World in Data*, 2015. See: https://ourworldindata.org/life-expectancy.

173 *Ray Kurzweil:* Ray Kurzweil, author interview, 2018. See also this conversation between Peter and Ray where they discuss the concept of longevity escape velocity: https://www.youtube.com/watch?time_continue=2&v=SaOfLtoaKqw.

173 *Aubrey de Grey:* Kira Peikoff, "Anti-Aging Pioneer Aubrey de Grey: 'People in Middle Age Now Have a Fair Chance,'" *Leapsmag*, January 30, 2018. See: https://leapsmag.com/anti-aging-pioneer-aubrey-de-grey-people-middle-age-now-fair-chance/.

The Anti-Aging Pharmacy

174 *Easter Island is remote: Joe Schwarz, "The Right Chemistry: Easter Island Might Just Hold the Key to Fighting Aging," Montreal Gazette, March 5, 2019.*

174 *rapamycin:* Bethany Halford, "Rapamycin's Secrets Unearthed," *C&EN*, July 18, 2016. See also: Bill Gifford, "Does a Real Anti-Aging Pill Already Exist?" *Bloomberg*, July 12, 2015.

175 *coating heart stents:* P. W. Serruys, "Rapamycin Eluting Stent: The Onset of a New Era in Interventional Cardiology," *Heart*, April 2002.

175 *don't reject their new kidney:* C. Morath, "Sirolimus in Renal Transplantation," *Nephrology Dialysis Transplantation*, September 2007.

175 *Rapamycin inhibits cancer growth:* Yekaterina Y. Zaytseva, "mTOR Inhibitors in Cancer Therapy," *Cancer Letters*, June 2012, doi.org/10.1016/j.canlet.2012.01.005.

175 *rapamycin extends the lives of mice by as much as 16 percent:* David E. Harrison, "Rapamycin Fed Late in Life Extends Lifespan in Genetically Heterogeneous Mice," *Nature*, July 8, 2019.

175 *Novartis to decide to test it in humans:* J. B. Mannick, "mTOR Inhibition Improves Immune Function in the Elderly," *Science Translational Medicine*, December 2014.

175 *drug called metformin:* Nir Barzilai, "Metformin as a Tool to Target Aging," *Cell Metabolism*, June 14, 2016. 23(6): pp. 1060-1065, See https://www.ncbi.nlm.nih.gov/pmc/articles/PMC5943638/.

176 *Jeff Bezos, Paul Allen, and Peter Thiel, Unity Biotechnology:* See: https://unitybiotechnology.com.

176 *extend lifespan by 35 percent:* Jan M van Deursen, "Senolytic Therapies for Healthy Longevity," *Science* 364, no. 6441 (May 2019): 636–637.

176 *Samumed:* Osman Kibar, author interview, 2018. See also: https://www.samumed.com/default.aspx.

176 *Backed by a $12 billion valuation:* Brian Gormley, "Drugmaker Samumed Closes $438 Million Round at $12 Billion Pre-Money Valuation," *Wall Street Journal Pro*, August 6, 2018. See: https://www.wsj.com/articles/drugmaker-samumed-closes-438-million-round-at-12-billion-pre-money-valuation-1533602014.

176 *one way the body sends messages*: Michael Khan, "Wnt Signaling in Stem Cells and Cancer Stem Cells: A Tale of Two Coactivators," *Science Direct*, January 2018. For an overview of the Wnt signaling pathway, see: https://www.sciencedirect.com/science/article/pii/S1877117317301850.

176 *twenty different diseases:* Kibar, author interview.

176 *Samumed has developed nine different so-called "regenerative medicines"*: See: https://www.samumed.com/pipeline/default.aspx.

176 *350 million people worldwide:* Juyoung Park, "Various Types of Arthritis in the United States: Prevalence and Age-Related Trends from 1999 to 2014," *American Journal of Public Health*, October 5, 2017. While this study only considers the US, the lead researcher, Juyoung Park, PhD, later extrapolated the study's result here: https://www.fau.edu/newsdesk/articles/arthritis-trends.php.

176 *results of a small study on knee osteoarthritis:* Y Yazici, "A Novel Wnt Pathway

Inhibitor, SM04690, for the Treatment of Moderate to Severe Osteoarthritis of the Knee: Results of a 24-Week, Randomized, Controlled, Phase 1 Study," *Osteoarthritis and Cartilage*, October 2017. See: https://www.ncbi.nlm.nih.gov/pubmed/28711582.

177 *Samumed CEO Osman Kibar:* Kibar, author interview.

177 *repair both rotator cuff and Achilles tendon injuries:* Samumed LLC, "A Study Evaluating the Safety, Tolerability, and Pharmacokinetics of Multiple Ascending Doses of SM04755 Following Topical Administration to Healthy Subjects," *U.S. National Library of Medicine*, July 25, 2017. See also: Samumed LLC, "A Repeat Insult Patch Test (RIPT) Study Evaluating the Sensitization Potential of Topical SM04755 Solution in Healthy Volunteers," *U.S. National Library of Medicine*, April 18, 2018. See: https://clinicaltrials.gov/ct2/show/NCT03229291?term=samumed&recrs=e&phase=0&rank=1. See also: https://clinicaltrials.gov/ct2/show/NCT03502434?term=samumed&recrs=e&phase=0&rank=2.

177 *knee arthritis drug is now entering Phase III:* Samumed LLC, "A Study Utilizing Patient-Reported and Radiographic Outcomes and Evaluating the Safety and Efficacy of Lorecivivint (SM04690) for the Treatment of Moderately to Severely Symptomatic Knee Osteoarthritis (STRIDES-X-ray)," *U.S. National Library of Medicine*, April 26, 2019. See: https://clinicaltrials.gov/ct2/show/NCT03928184?term=samumed&phase=2&rank=1.

177 *halted tumor growth in 80 percent of the study group:* Kibar, author interview.

The Bloody Fountain of Youth

178 *a group of Stanford researchers:* D. E. Wright, "Physiological Migration of Hematopoietic Stem and Progenitor Cells," *Science*, November 2001. See: https://www.ncbi.nlm.nih.gov/pubmed/11729320.

178 *rejuvenating effects of young blood:* Megan Scudellari, "Ageing Research: Blood to Blood," *Nature*, January 21, 2015. See: https://www.nature.com/news/ageing-research-blood-to-blood-1.16762. note for citation: using the magazine info at this link may help: https://static1.squarespace.com/static/5b7168984eddec1ee06cba31/t/5b7cfbc3aa4a99594f9ff075/1534917589733/Nature.pdf).

178 *the formation of new neurons in the brain:* L. Katsimpardi, "Vascular and Neurogenic Rejuvenation of the Aging Mouse Brain by Young Systemic Factors," *Science*, May 2014. See: https://www.ncbi.nlm.nih.gov/pubmed/24797482.

178 *walls of the heart:* Francesco S. Loffredo, "Growth Differentiation Factor 11 Is Circulating Factor That Reverses Age-Related Cardiac Hypertrophy," *Cell*, May 9, 2013. See: https://www.cell.com/abstract/S0092-8674(13)00456-X.

178 *2014 paper published in Cell:* Ibid.

179 *Elevian:* See: https://www.elevian.com/.

179 *Mark Allen:* Mark Allen, author interview, 2018.

179 *Alkahest:* See: https://www.alkahest.com/.

179 Wired *called these kinds of efforts:* Megan Molteni, "Startups Flock to Turn Young Blood into an Elixir of Youth," *Wired*, September 5, 2018. See: https://www.wired.com/story/startups-flock-to-turn-young-blood-into-an-elixir-of-youth/.

179 *National Institute on Aging committed $2.35 million:* Ibid.

Chapter Eleven: The Future of Future of Insurance, Finances, and Real Estate

Coffee, Risk, and the Origins of Insurance

182 *Edward Lloyd:* For a full history of Lloyd's of London, see: https://www.lloyds.com/about-lloyds/history/corporate-history.

183 *33.6 billion pounds' worth of insurance premiums in 2017:* Read the full press release from March 27, 2019, here: https://www.lloyds.com/news-and-risk-insight/press-releases/2019/03/lloyds-reports-aggregated-market-results-for-2018.

The Car That Doesn't Crash

184 *Car insurance premiums:* The Insurance Information Institute lays out what determines your car insurance premiums here: iii.org/article/what-determines-price-my-auto-insurance-policy.

184 *90 percent:* Read the full "Critical Reasons for Crashes Investigated in the National Motor Vehicle Crash Causation Survey" report from the National Highway Traffic Safety Administration here: https://crashstats.nhtsa.dot.gov/Api/Public/ViewPublication/812115.

184 *1.2 million traffic fatalities a year:* This data comes from the World Health Organization, "Road Traffic Deaths Data by Country," found here: http://apps.who.int/gho/data/node.main.A997.

184 *KPMG predicts the car insurance market:* Download KPMG's "Road Traffic Injuries and Deaths—A Global Problem," report from June 2015 here: https://www.insurancejournal.com/research/research/kpmg-automobile-insurance-in-the-era-of-autonomous-vehicles/.

184 *10 million miles:* John Krafcik, "Where the Next 10 Million Miles Will Take Us," *Medium,* October 10, 2018. See: https://medium.com/waymo/where-the-next-10-million-miles-will-take-us-de51bebb67d3.

185 *1 million people are arrested for DUIs each year:* See the FBI's statistics on 2016 crime rates in the US here: https://ucr.fbi.gov/crime-in-the-u.s/2016/crime-in-the-u.s.-2016/tables/table-18.

Crowdsurance

186 *Lemonade:* See: https://www.lemonade.com/.

187 *Etherisc:* See: https://etherisc.com/.

187 *flight delays and cancellations:* Etherisc, "First Blockchain-Based App to Insure Your Next Flight Against Delays," *Medium*, July 23, 2018. See: https://blog.etherisc.com/first-blockchain-based-app-to-insure-your-next-flight-against-delays-10f53b38ad2d.

Dynamic Risk

187 *Progressive Insurance:* Read more about the company's history here: https://www.progressivecommercial.com/about-us/our-history/.

187 *first insurance company to have a website:* See Progressive's long list of firsts here: https://www.progressive.com/about/firsts/.

188 *"TripSense":* Read the full press release announcing TripSense here: https://www.businesswire.com/news/home/20040809005574/en/Innovative-Auto-Insurance-Discount-Program-5000-Minnesotans.

188 *30 percent of all home claims are from water damage:* According to the Travelers Insurance Company, 20 percent of their home insurance claims are from non-weather-related water damage, while 11 percent are from weather-related water damage. See: https://www.travelers.com/tools-resources/home/maintenance/top-five-ways-things-can-go-wrong-interactive.

189 *The term McKinsey coined:* McKinsey & Co., "Digital Insurance in 2018: December 2018 Driving Real Impact with Digital and Analytics," December 2018. See: https://www.mckinsey.com/~/media/McKinsey/Industries/Financial%20Services/Our%20Insights/Digital%20insurance%20in%202018%20Driving%20real%20impact%20with%20digital%20and%20analytics/Digital-insurance-in-2018.ashx.

189 *will drop by 70 to 90 percent:* Ibid., p. 38.

Good Money

190 *Gunnar Lovelace:* Gunnar Lovelace, author interview, 2018. For more information on Gunnar Lovelace, see his LinkedIn bio: https://www.linkedin.com/in/gunnarlovelace/.

190 *Thrive Market:* See: https://thrivemarket.com.

190 *Good Money:* See: https://goodmoney.com.

190 *On average, we pay $360 a year in banking fees:* Lovelace, author interview.

190 *$30 billion a year in overdraft:* Maria Lamagna, "Overdraft Fees Haven't Been This Bad Since the Great Recession," *MarketWatch*, April 2, 2018. See: https://www.marketwatch.com/story/overdraft-fees-havent-been-this-bad-since-the-great-recession-2018-03-27.

190 *lost a ton of business:* Seattle, Washington, and Davis, California, together pulled more than $3 billion from Wells Fargo: Bill Chappell, "2 Cities to Pull More Than $3 Billion from Wells Fargo Over Dakota Access Pipeline," NPR, February 8, 2017. See: https://www.npr.org/sections/thetwo-way/2017/02/08/514133514/two-cities-vote-to-pull-more-than-3-billion-from-wells-fargo-over-dakota-pipeline.

191 *while over two billion people in the world still lack bank accounts:* Niall McCarthy, "1.7 Billion Adults Worldwide Do Not Have Access to a Bank Account," *Forbes*, June 8, 2018. See original data here: https://globalfindex.worldbank.org.

An Unusual Proposition

191 *Vodafone's Nick Hughes:* For a comprehensive look at the MPESA story, see Tim Harford, *Fifty Inventions That Shaped the Modern Economy* (Penguin, 2017), p. 228.

192 *The result of this collaboration was M-Pesa:* Ibid. p. 229.

192 *12 percent:* Western union fees range from 2 percent to 30 percent depending on where you are and how much you're transferring. See their fee table here: https://www.westernunion.com/content/dam/wu/EU/EN/feeTableRetailEN-ES.PDF.

192 *Eight months after launch:* Harford, *Fifty Inventions*, p. 229.

192 *According to research done at MIT:* Tavneet Suri, "The Long-Run Poverty and Gender Impacts of Mobile Money," *Science* 354, no. 6317 (December 9, 2016): 1288–1292. See: https://science.sciencemag.org/content/354/6317/1288.

192 *30 million people in ten different countries:* Read the feature news story "What Kenya's Mobile Money Success Could Mean for the Arab World" on the World Bank's website here: https://www.worldbank.org/en/news/feature/2018/10/03/what-kenya-s-mobile-money-success-could-mean-for-the-arab-world.

192 *bKash now serves over 23 million users:* Read the International Finance Corporation's Inclusive Business Case Study of bKash here: http://documents.worldbank.org/curated/en/560181506580665929/pdf/119870-BRI-PUBLIC-bKash-Builtforchangereport.pdf.

192 *Alipay serves just shy of a billion:* Xinhua, "China's Alipay Now Has Over 900m Users Worldwide," *China Daily*, November 30, 2018. See: http://www.chinadaily.com.cn/a/201811/30/WS5c00a1d3a310eff30328c073.html.

192 *500 million customers play "Ant Forest":* Christine Chou, "How Alipay Users Planted 100M Trees in China," *Alizila*, April 22, 2019. See: https://www.alizila.com/how-alipay-users-planted-100m-trees-in-china/.

193 *R3:* See: https://www.r3.com/.

193 *Ripple:* See: https://www.ripple.com/.

193 *these companies are using blockchain to replace the SWIFT network:* The blockchain blog *Cointelegraph* has a good piece overviewing SWIFT's relationship to blockchain. You can find the piece here: https://cointelegraph.com/news/swift-announces-poc-gateway-with-r3-but-remains-overall-hesitant-about-blockchain.

193 *4 billion people, the rising billions, will gain access to the internet:* The world population is expected to reach 8.2 billion by 2025, as projected by the United Nations Population Division. You can find the raw data here: https://population.un.org/wpp/Graphs/Probabilistic/POP/TOT/900. On the other side of the equation, we use data from the World Bank to approximate how many people are connected to the internet right now; World Bank, "Individuals using the Internet (% of population)." See: https://data.worldbank.org/indicator/IT.net.USER.ZS See also: "Population, total," The World Bank, https://data.worldbank.org/indicator/SP.POP.TOTL.

The AI Invasion

194 *TransferWise:* See: https://transferwise.com/us.

194 *Prosper:* See: https://www.prosper.com.

194 *Funding Circle:* See: https://www.fundingcircle.com/us/.

194 *LendingTree:* See: https://www.lendingtree.com/.

194 *$26.16 billion in 2015 to $897.85 billion by 2024:* According to a report by Transparency Market Research, found here: https://www.globenewswire.com/news-release/2016/08/31/868470/0/en/Increasing-Small-Business-Units-to-Act-as-Building-Blocks-for-Peer-to-Peer-Lending-Market.html.

194 *Smart Finance Group:* Alexandra Stevenson, "China's New Lenders Collect Invasive Data and Offer Billions. Beijing Is Worried," *New York Times*, December 25, 2017. See: https://www.nytimes.com/2017/12/25/business/china-online-lending-debt.html.

195 *Wealthfront:* See: https://www.wealthfront.com/.

195 *Betterment:* See: https://www.betterment.com/.

195 *60 percent of all market trades:* Chris Isidore, "Machines Are Driving Wall Street's Wild Ride, Not Humans," CNN, February 6, 2018. See: https://money.cnn.com/2018/02/06/investing/wall-street-computers-program-trading/index.html.

195 *90 percent:* Ibid.

195 *robo-advisors take around .25 percent:* See Wealthfront and Betterment's websites for their rates.

195 *Wealthfront had $11 billion under management:* Read the full SEC press release, titled "SEC Charges Two Robo-Advisers with False Disclosures," here: https://www.sec.gov/news/press-release/2018-300.

195 *Betterment was at $14 billion:* Garrett Keyes, "How Betterment Stayed on Top in 2018 (and How They Plan to Stay There in 2019)," *Financial Advisor IQ*, January 2, 2019. See: https://www.betterment.com/press/newsroom/how-betterment-stayed-on-top-in-2018-and-how-they-plan-to-stay-there-in-2019/.

195 Business Insider Intelligence *estimates:* Sarah Kocianski, "The Evolution of Robo-Advising: How Automated Investment Products Are Disrupting and Enhancing the Wealth Management Industry," *Business Insider*, July 3, 2017. See: https://www.businessinsider.com/the-evolution-of-robo-advising-report-2017-7.

196 *Uber Eats:* Uber Eats is one of a handful of app-based food delivery startups operating around the world. See their website here: http://ubereats.com.

196 *Denmark stopped printing money in 2017:* Peter Levring, "Scandinavia's Disappearing Cash Act," *Bloomberg*, December 15, 2016. See: https://www.bloomberg.com/news/articles/2016-12-16/scandinavia-s-disappearing-cash-act.

196 *India recalled 86 percent of its cash:* Read UBS's full report, "The Road to Cashless Societies: Shifting Asia," here: https://www.ubs.com/content/dam/WealthManagementAmericas/cio-impact/shifting%20in%20asia.pdf. This specific fact comes from p. 19.

196 *Vietnam wants retail to be 90 percent cashless by 2020:* "Cashless Payment Posts Double-Digit Growth," *Viet Nam News*, July 13, 2019. See: https://vietnamnews.vn/economy/522587/cashless-payment-posts-double-digit-growth.html#kpUKGbUeSj1J1IzH.97.

196 *Sweden, where over 80 percent of all transactions are digital:* Read more about this topic on Sweden's official website: https://sweden.se/business/cashless-society/.

Real Estate

196 *Great Recession of 2008:* Michael Lewis gives a detailed look at what caused the 2008 financial crisis in his *New York Times* bestselling book *The Big Short: Inside the Doomsday Machine* (W. W. Norton & Company, 2011).

196 *Glenn Sanford:* Glenn Sanford, author interview, 2019.

196 *eXp Realty:* See: https://www.exprealty.com.

196 *which eXp now owns:* Ibid.

197 *sixteen thousand agents:* Ibid.

197 *$100 million of his company's $650 million market cap:* Ibid.

Say Goodbye to Your Broker

197 *Zillow:* See: https://www.zillow.com/. You can learn more about Zillow's AI strategy in this interview with Zillow's chief analytics officer, Stan Humphries: Michael Krigsman, "Zillow: Machine Learning and Data Disrupt Real Estate," ZDNet, July 30, 2017, https://www.zdnet.com/article/zillow-machine-learning-and-data-in-real-estate/.

197 *Trulia:* See: https://www.trulia.com/.

197 *Move:* See: https://www.move.com/.

197 *Redfin:* See: https://www.redfin.com/.

197 *invested millions:* For an example of investment in real estate AI, see this *VentureBeat* article about one of Zillow's latest computer vision tools: Kyle Wiggers, "Zillow Now Uses Computer Vision To Improve Property Value Estimates," *VentureBeat,* June 26, 2019. See: https://venturebeat.com/2019/06/26/zillow-now-uses-computer-vision-to-improve-property-value-estimates/.

Reinventing the City

199 *five hundred coastal cities now threatened by global warming:* This World Economic forum report predicts that 570 coastal cities around the world are vulnerable to a sea-level rise of 0.5 meters by 2050: http://www3.weforum.org/docs/WEF_Global_Risks_Report_2019.pdf.

199 *close to 40 percent of humanity already lives near the ocean:* The UN reports that nearly 2.4 billion people (about 40 percent of the world's population) live within 100 km (60 miles) of the coast. See the UN fact sheet here: https://www.un.org/sustainabledevelopment/wp-content/uploads/2017/05/Ocean-fact-sheet-package.pdf.

200 *Oceanix City:* See: http://oceanix.org/. For a profile on the company, see also: Katharine Schwab, "Floating Cities Once Seemed Like Sci-Fi. Now the UN Is Getting On Board," *Fast Company,* https://www.fastcompany.com/90329294/floating-cities-once-seemed-like-sci-fi-now-the-un-is-getting-on-board.

200 *Seasteading Institute:* See: https://www.seasteading.org/.

200 *tested in the waters off French Polynesia:* David Gelles, "Floating Cities, No Longer Science Fiction, Begin to Take Shape," *New York Times,* November 13, 2017. See: https://www.nytimes.com/2017/11/13/business/dealbook/seasteading-floating-cities.html.

Chapter Twelve: The Future of Food

201 *3-D food printer:* Jonathan Chadwick, "Here's How 3-D Printers Are Changing What We Eat," *TechRepublic,* November 7, 2017. See: https://

www.techrepublic.com/article/heres-how-3-D-food-printers-are-changing-the-way-we-cook/.

201 *the robot arm slicing fresh tuna with a smooth motion:* Stuart Farrmiond, "The Future of Food: What We'll Eat in 2028," *Science Focus*, May 17, 2019. See: https://www.sciencefocus.com/future-technology/the-future-of-food-what-well-eat-in-2028/.

202 *lab-grown provider:* Matt Simon, "Lab Grown Meat Is Coming, Whether You Like It or Not," *Wired*, February 16, 2018. See: https://www.wired.com/story/lab-grown-meat/.

The Inefficiency of Food

202 *author Richard Manning wrote in an essay for* Harpers*:* Richard Manning, "The Oil We Eat: Following the Food Chain Back to Iraq," *Harpers*, February 2004. See: https://harpers.org/archive/2004/02/the-oil-we-eat/.

203 *The average American meal travels:* Rich Pirog, "The Evolution of Food Miles and Its Limitations as an Indicator of Energy Use and Climate Impact." See: https://aceee.org/files/pdf/conferences/ag/2008/RPirog.pdf.

203 *according to the National Resources Defense Council:* Move for Hunger, "About Food Waste." See: https://www.moveforhunger.org/about-food-waste/.

203 *researchers at UCLA:* Amina Khan, "Scientists Aim to Feed the World by Boosting Photosynthesis," *Los Angeles Times*, November 18, 2016. See: https://www.latimes.com/science/sciencenow/la-sci-sn-boosting-photosynthesis-20161117-story.html.

203 *The Bill Gates–backed RIPE Project at the University of Illinois:* Claire Benjamin, "Scientists Boost Crop Production by 47 Percent Speeding Up Photorespiration," *Ripe Project*, May 31, 2018. See: https://ripe.illinois.edu/press/press-releases/scientists-boost-crop-production-47-percent-speeding-photorespiration.

203 *The UN estimates that we need to double:* "Food Production Must Double by 2050 to Meet Demand from World's Growing Population, Innovative Strategies Needed to Combat Hunger, Experts Tell Second Committee," *United Nations*, October 9, 2009. https://www.un.org/press/en/2009/gaef3242.doc.htmfood.

203 *Santa Barbara–based Apeel Sciences:* See: https://apeelsciences.com/.

204 *Avocados:* "Apeel Avocados Expected at Every U.S. Grocery Store Within a Year," Freshfruitportal.com, October 24, 2018. See: https://www.freshfruitportal.com/news/2018/10/24/apeel-avocados-expected-at-every-u-s-grocery-store-within-a-year/.

204 *idea known as "vertical farming":* Dickson Despommier, *The Vertical Farm* (Picador, 2011).

204 *45 percent:* Caryn Roni Rabin, "Do Prepackaged Salad Greens Lose Their

Nutrients?" *New York Times,* November 3, 2017. See: https://www.nytimes.com/2017/11/03/well/eat/do-prepackaged-salad-greens-lose-their-nutrients.html.

204 *Relying on hydroponics and aeroponics:* Despommier, "Vertical Farm." See also: Lisa Grace Scott, "Vertical Garden Towers Can Grow Plants Three Times Faster Than Normal: How a Business in the Bronx Is Trying to Take Urban Gardening Mainstream," *Inverse*, June 1, 2018, https://www.inverse.com/article/45464-rooftop-garden-technology-vertical-garden.

204 *Bay Area–based Plenty Unlimited Inc.:* Dee: https://www.plenty.ag/.

204 *With over $200 million in funding:* Chelsea Ballarte, "Jeff Bezos and Other Investors Raise $200 Million for Vertical Farming Startup Plenty," *GeekWire*, July 20, 2017. See: https://www.geekwire.com/2017/jeff-bezos-investors-raise-200-million-vertical-farming-startup-plenty/.

205 *seventy-thousand-square-foot factory:* Olivia Solon, "Inside the World's Largest Vertical Farm," *Wired*, February 29, 2016. See: https://www.wired.co.uk/article/aerofarms-largest-vertical-farm.

205 *AeroFarms:* See: https://aerofarms.com/.

205 *as Plenty CEO Matt Barnard recently told reporters:* "Agtech Startup Plenty Plans To Grow Hydroponic Peaches," Matteroftrust.org. See: https://matteroftrust.org/agtech-start-up-plenty-plans-to-grow-hydroponic-peaches/.

205 *50 to 80 percent of a vertical farm's cost is human labor:* Andrew Tarantola, "The Future of Indoor Agriculture Is Indoor Farms Run by Robots," *engadget*, October 3, 2018. See: https://www.engadget.com/2018/10/03/future-indoor-agriculture-vertical-farms-robots/.

205 *Silicon Valley–based Iron Ox:* See: http://ironox.com/.

205 *as* Engadget *recently wrote:* Ibid.

The Inefficiency of Growing a Cow

206 *world will need 70 percent more food than it did in 2009:* Food and Agriculture Organization of the United Nations, "2050: A Third More Mouths to Feed," *FAO News,* September 3, 2009. See: http://www.fao.org/news/story/en/item/35571/icode/.

206 *global meat consumption is expected to increase by 76 percent:* United Nations, "What's in Your Burger? More Than You Think," *UN Environment,* November 8, 2018. See: http://www.unenvironment.org/news-and-stories/story/whats-your-burger-more-you-think.

206 *50 percent of all habitable land on Earth is used for agriculture:* Natasha Brooks, "Chart Shows What the World's Land Is Used for . . . and It Explains Exactly Why So Many People Are Going Hungry," *One Green Planet*, 2018. See: https://www.onegreenplanet.org/news/chart-shows-worlds-land-used/.

206 *A quarter of the planet's available landmass is currently used to:* Timothy P.

Robinson, "Mapping the Global Distribution of Livestock," *PLoS ONE*, May 29, 2014. See: https://journals.plos.org/plosone/article?id=10.1371/journal.pone.0096084#pone-0096084-g002. See also: "Counting Chickens," *Economist*, July 27, 2011, https://www.economist.com/graphic-detail/2011/07/27/counting-chickens.

206 *Meat production accounts for 70 percent of global water:* "Water for Sustainable Food and Agriculture: A Report Produced for the G20 Presidency of Germany," *Food and Agriculture Organization of the United Nations*, 2017. See: http://www.fao.org/3/a-i7959e.pdf.

206 *fifteen hundred liters required to:* Charles Ebikeme, "Water World," *Scitable*, July 25, 2013. See: https://www.nature.com/scitable/blog/eyes-on-environment/water_world/.

206 *Meat is also responsible for 14.5 percent of all greenhouse gases:* Lisa Friedman, Kendra Pierre-Louis, and Somini Sengupta, "The Meat Question, by the Numbers," *New York Times*, January 25, 2018. See: https://www.nytimes.com-2018/01/25-climate-cows-global-warming.

206 *loss of land for agricultural production is currently the largest driver of that extinction:* Damian Carrington, "Avoiding Meat and Dairy Is 'Single Biggest Way' to Reduce Your Impact on Earth," *Guardian*, May 31, 2018. See: https://www.theguardian.com/environment/2018/may/31/avoiding-meat-and-dairy-is-single-biggest-way-to-reduce-your-impact-on-earth.

206 *the cultured meat formula:* Sam Baker, "Will 2019 Be the Year of Lab Grown Meat?" DW.com, March 1, 2019. See https://www.dw.com/en/will-2019-be-the-year-of-lab-grown-meat/a-46943665.

207 *Cultured meat uses:* Hanna L. Tuomisto and Joost M. Teixeira de Mattos, "Environmental Impacts of Cultured Meat Production," *Environmental Science and Technology*, June 17, 2011. See: https://doi.org/10.1021/es200130u and https://pubs.acs.org/doi/10.1021/es200130u.

207 *Cultured meat is also a healthier solution:* Marta Zaraska, "Is Lab Grown Meat Good for Us?" *Atlantic*, August 19, 2013. See: https://www.theatlantic.com/health/archive/2013/08/is-lab-grown-meat-good-for-us/278778/.

207 *70 percent of emerging diseases come from livestock:* Food and Agriculture Organization of the United Nations, "Surge in Diseases of Animal Origin Necessitates New Approach to Health," *FAO*, December 16, 2013. See: http://www.fao.org/news/story/en/item/210621/icode/.

207 *Back in 2013, the first cultured burger cost $330,000:* Pallab Ghosh, "World's First Lab-Grown Burger Is Eaten in London," *BBC News*, August 5, 2013. See: https://www.bbc.com/news/science-environment-23576143.

208 *Memphis Meats:* Chloe Sorvino, "Tyson Invests in Lab-Grown Protein Startup Memphis Meats, Joining Bill Gates and Richard Branson," *Forbes*,

January 29, 2018. See: https://www.forbes.com/sites/chloesorvino/2018/01/29/exclusive-interview-tyson-invests-in-lab-grown-protein-startup-memphis-meats-joining-bill-gates-and-richard-branson/#5f4025763351.

208 *Aleph Farms has steak down to around $50 a pound:* Leanne Back and Tava Cohen, "On the Menu Soon: Lab-Grown Steak for Eco-Conscious Diners," Reuters, July 15, 2019. See: https://www.reuters.com/article/us-food-tech-labmeat-aleph-farms/on-the-menu-soon-lab-grown-steak-for-eco-conscious-diners-idUSKCN1UA1ES.

208 *cultured meat has the potential:* Yaakov Nahmias, "Lab-Grown Meat Is Getting Cheap Enough for Anyone to Buy," *Fast Company*, May 2, 2018. See: https://www.fastcompany.com/40565582/lab-grown-meat-is-getting-cheap-enough-for-anyone-to-buy.

208 *Just Inc. announced a partnership with Japanese Wagyu beef producer:* Adele Peters, "The Meat Growing in This San Francisco Lab Will Soon Be Available at Restaurants," *Fast Company*, December 11, 2018. See: https://www.fastcompany.com/90278853/the-meat-growing-in-this-san-francisco-lab-will-soon-be-available-at-restaurants.

208 *Perfect Day Foods:* See: https://www.perfectdayfoods.com/. See also: Alexandra Wilson, "Got Milk? This $40M Startup Is Creating Cow-Free Dairy Products That Taste like the Real Thing," *Forbes*, January 9, 2019.

PART 3: THE FASTER FUTURE

Chapter Thirteen: Threats and Solutions

Water Woes

212 *"Special Report on Global Warming":* The intergovernmental panel on climate change, see: https://www.ipcc.ch/sr15/.

212 *Global Risks Report:* World Economic Forum, "Global Risks Report 2018: 13th Edition," January 17, 2018. See: https://www.weforum.org/reports/the-global-risks-report-2018.

213 *Dean Kamen:* Dean Kamen, author interview, 2018. For more information about Dean Kamen, see his bio on the FIRST Robotics website here: https://www.firstinspires.org/about/leadership/dean-kamen.

213 *we recounted the story of Kamen's "Slingshot":* Peter Diamandis and Steven Kotler, *Abundance: The Future Is Better Than You Think* (Free Press, 2012), pp. 88–91.

213 *900 million people lack access to clean drinking water:* World Health Organization (WHO) and the United Nations Children's Fund (UNICEF),

"Progress on Drinking Water, Sanitation and Hygiene," 2017. See: https://apps.who.int/iris/bitstream/handle/10665/258617/9789241512893-eng.pdf;jsessionid=1FDA500FD803F836724FE17B699EE7AA?sequence=1.

213 *3.4 million lives a year:* United Nations Educational, Scientific, and Cultural Organization, "Managing Water Report Under Uncertainty And Risk—the United Nations World Water Development Report 4 Volume 1," 2012, p. 96. See: http://www.unesco.org/new/fileadmin/MULTIMEDIA/HQ/SC/pdf/WWDR4%20Volume%201-Managing%20Water%20under%20Uncertainty%20and%20Risk.pdf.

213 *half the globe will be water stressed:* World Health Organization, "Drinking Water," June 14, 2019. See: https://www.who.int/news-room/fact-sheets/detail/drinking-water.

213 *Kamen had just made a handshake deal with Coca-Cola:* Coca Cola Corporate, "Coca-Cola Announces Long-Term Partnership with DEKA R&D to Help Bring Clean Water to Communities in Need," September 25, 2012. See Coca-Cola's press release about their agreement here: https://www.coca-colacompany.com/press-center/press-releases/deka-partnership-announcement. See also: "EKOCENTER & Slingshot Clean Water Partnerships," https://www.coca-colaafrica.com/stories/sustainability-water-ekocenter#.

214 *"Freestyle Fountain Beverage Dispenser":* Ted Ryan, "From Big Idea to Big Bet: How the Coca-Cola Freestyle Fountain Dispenser Came to Be," December 6, 2017. See: https://www.coca-colacompany.com/stories/freestyle-q-a.

214 *"Ekocenter":* Ibid. See: https://www.coca-colaafrica.com/stories/sustainability-water-ekocenter#.

214 *By 2017, there were 150 Ekocenters operating in eight countries:* The Coca Cola Company, "Scaling Sustainability: Have Programs, Will Travel," August 24, 2018. See: https://www.coca-colacompany.com/stories/sustainability-lift-and-shift-have-programs-will-travel.

214 *78.1 million liters:* Ibid.

214 *Skysource:* See: http://www.skysource.org/.

214 *$1.5 million Water Abundance XPRIZE:* See: https://www.xprize.org/prizes/water-abundance. See also: Devin Coldewey, "Water Abundance Xprize's $1.5M Winner Shows How to Source Fresh Water from the Air," TechCrunch, October 22, 2018. See: https://techcrunch.com/2018/10/22/water-abundance-xprizes-1-5m-winner-shows-how-to-source-fresh-water-from-the-air/.

214 *350 and 400 million gallons a day:* Adele Peters, "A Device That Can Pull Drinking Water from the Air Just Won the Latest XPrize" *Fast Company,*

October 20, 2018. See: https://www.fastcompany.com/90253718/a-device-that-can-pull-drinking-water-from-the-air-just-won-the-latest-x-prize.

214 *"smart grid for water"*: Trevor Hill, *The Smart Grid for Water: How Data Will Save Our Water and Your Utility* (Advantage, 2013).

215 *saving us trillions of gallons a year:* Ibid. In the US alone, we lose at least an estimated 1.7 trillion gallons of water per year to water main breaks.

Climate Change for Optimists

215 *Forty billion tons of CO_2:* According to Maxwell Rosner's *Our World in Data*, 35.46 billion tons of CO_2 were emitted in 2017. See: https://ourworldindata.org/co2-and-other-greenhouse-gas-emissions.

215 *Caleb Scharf tried to find a comparison:* Caleb Scharf, "The Crazy Scale of Human Carbon Emission," *Scientific American*, April 26, 2017. See: https://blogs.scientificamerican.com/life-unbounded/the-crazy-scale-of-human-carbon-emission/.

215 *Carbon Majors Database:* Paul Griffin, "CDP Carbon Majors Report, 2017," *Carbon Majors Database,* July 2017. See: https://b8f65cb373b1b7b15feb-c70d8ead6ced550b4d987d7c03fcdd1d.ssl.cf3.rackcdn.com/cms/reports/documents/000/002/327/original/Carbon-Majors-Report-2017.pdf?1499691240.

216 *For comparison purposes:* Energy prices discussed here are derived from an author interview with Ramez Naam, head of Energy, Climate and Innovation for Singularity University, 2019. For the most part, the cost of energy data points in this section of the book use LCOE (levelized cost of energy) estimates obtained from the U.S. National Renewable Energy Laboratory's openei.org Transparent Cost Database. See : https://openei.org/apps/TCDB/.

216 *6 cents a kilowatt-hour:* Ibid.

216 *In the 1980s, the energy produced by a new wind plant cost 57 cents a kilowatt-hour:* Ibid.

216 *it's 2.1 cents (if you remove all subsidies, it's 4 cents):* Ibid.

216 *That's a 94 percent decrease in price:* Ibid.

216 *bringing us "one cent wind" by 2030:* Ibid.

216 *This price-performance curve is like nothing we've ever seen in energy:* Ibid.

216 *75 to 90 percent as eight of America's largest coal companies:* These include: James River Coal Company (2014), Patriot Coal (2015), Walter Energy (2015), Alpha Natural Resources (2015), Peabody Energy (2016), Blackhawk Mining (2019), Blackjewel (2019), Cloud Peak Energy (2019).

216 *China canceled the construction of 160 plants:* Ibid. See also: Michael Forsythe, "China Cancels 103 Coal Plants, Mindful of Smog and Wasted

Capacity," *New York Times*, January 18, 2017. See: https://www.nytimes.com/2017/01/18/world/asia/china-coal-power-plants-pollution.html.

216 *killing $9 billion:* Ibid.

217 *recently became a solar farm:* Ontario Power Generation's website has more details about their Nanticoke Generating Station solar farm project here: https://www.opg.com/strengthening-the-economy/our-projects/nanticoke-solar-facility/.

217 *more energy from zero-carbon sources than coal:* Jillian Ambrose, "Fossil Fuels Produce Less Than Half of UK Electricity for First Time," *Guardian*, June 20, 2019. See: https://www.theguardian.com/business/2019/jun/21/zero-carbon-energy-overtakes-fossil-fuels-as-the-uks-largest-electricity-source.

217 *over a hundred major cities:* For a full list of the 100 plus major cities getting 70 percent of their energy from renewables, see the Carbon Disclosure Project's full list here: https://www.cdp.net/en/cities/world-renewable-energy-cities.

217 *8 percent of the world's electricity now comes from solar and wind:* See Figure 8 in Chapter 1 of the *REN21, Renewables 2019 Global Status Report*. Access the report here: https://www.ren21.net/gsr-2019/.

217 *costs less to build a new wind farm or solar plant:* The levelized cost of energy (LCOE) allows comparison of different methods of electricity generation. Looking at the U.S. National Renewable Energy Laboratory's Transparent Cost Database referenced above (https://openei.org/apps/TCDB/), you can see that wind and solar can be less expensive than coal.

217 *4.5 cents a kilowatt-hour:* Naam, author interview.

217 *3.8 cents:* Ibid. $0.038 reached for the 250 Megawatt Bhadla solar park project in 2017. See also: Mayank Aggarwal, "Solar Power Tariffs Fall to New Low of Rs2.62 Per Unit," *livemint*, 2017, https://www.livemint.com/Industry/MKI7QvOhpRoBAtw3-D4PM5K/South-African-firm-bid-takes-solar-power-tariffs-to-new-low.html.

217 *Abu Dhabi: 2.4 cents:* Naam, author interview. See also: Emiliano Bellini, "Dubai: Tariff for Large-Scale PV Hits New Low at $0.024/kWh," *PV Magazine*, 2018, https://www.pv-magazine.com/2018/11/05/dubai-tariff-for-large-scale-pv-hits-new-low-at-0-024-kwh/.

217 *Chile beat it with 2.1*: Naam, author interview.

217 *Brazil beat that with 1.75:* Naam, author interview.

217 *"quantum dots":* For background, see: Prashant Kamat, "Quantum Dot Solar Cells. The Next Big Thing in Photovoltaics," *Journal of Physical Chemistry*, February 28, 2013, pp. 908–918, https://doi.org/10.1021/jz400052e.

217 *21 percent of incoming sunlight:* Naam, author interview.

217 *66 percent:* Naam, author interview.

217 *two-thirds of solar's price comes from soft costs:* According to a National Renewable Energy Lab study from 2013, soft costs account for "64% of the total residential system price, 57% of the small (less than 250 kW) commercial system price, and 52% of the large (250 kW or larger) commercial system price." See: Barry Friedman, "Benchmarking Non-Hardware Balance-of-System (Soft) Costs for U.S. Photovoltaic Systems, Using a Bottom-Up Approach and Installer Survey—Second Edition," National Renewable Energy Laboratory, 2013.

218 *Every 88 minutes:* Naam, author interview.

The Story of Storage

218 *California decided to source 100 percent:* Jason Pontin, "We Gotta Get a Better Battery. But How?" *Wired*, September 17, 2018. See: https://www.wired.com/story/better-battery-renewable-energy-jason-pontin/.

219 *plummeting 90 percent between 1990 and 2010:* Naam, author interview.

219 *eleven-fold increase in capacity:* Ibid.

219 *Enter the Gigafactory:* See Gigafactory 1's website here: https://www.tesla.com/gigafactory. See also: Matthew Wald, "Nevada a Winner in Tesla's Battery Contest,"*New York Times*, September 4, 2014, https://www.nytimes.com/2014/09/05/business/energy-environment/nevada-a-winner-in-teslas-battery-contest.html.

219 *A second Gigafactory has been built in Buffalo:* See Gigafactory 2's website here: https://www.tesla.com/gigafactory2.

219 *a third in Shanghai:* Simon Alvarez, "China Formally Adds Tesla Gigafactory 3 Area to Shanghai's Free-Trade Zone," *Teslarati,* August 6, 2019. See: https://www.teslarati.com/china-adds-tesla-gigafactory-3-shanghai-free-trade-zone/.

219 *European location:* At the time that we are publishing this book, Tesla is still in the planning phase for Gigafactory 4. For more info, see this article: Simon Alvarez, "Tesla Closing In on Lower Saxony, Germany as Final Europe Gigafactory Location: Report," *Teslarati*, August 22, 2019, https://www.teslarati.com/tesla-europe-gigafactory-4-location-lower-saxony-germany/.

219 *one hundred Gigafactories:* Leonardo DiCaprio, *Before the Flood* (documentary), National Geographic, October 21, 2016. You can watch a relevant clip on Nat Geo's YouTube channel here: https://youtu.be/iZm_NohNm6I.

219 *Tesla built the largest battery facility ever:* Thuy Ong, "Elon Musk Has Finished Building the World's Biggest Battery in Less Than 100 Days," *Verge*, November 23, 2017. See: https://www.theverge.com/2017/11/23/16693848/elon-musk-worlds-biggest-battery-100-days.

219 *Zoe batteries:* See: https://www.renault.co.uk/vehicles/new-vehicles/zoe/motor.html.

219 *BMW's 500 i3 battery packs:* Jon Fingas, "BMW i3 Batteries Provide Energy Storage for UK Wind Farm," *Engadget*, May 21, 2018. See: https://www.engadget.com/2018/05/21/bmw-i3-battery-packs-join-uk-power-grid/.

219 *Flow batteries*: Robert Service, "New Generation of 'Flow Batteries' Could Eventually Sustain a Grid Powered by the Sun and Wind," *ScienceMag*, October 31, 2018. See: https://www.sciencemag.org/news/2018/10/new-generation-flow-batteries-could-eventually-sustain-grid-powered-sun-and-wind.

220 *handle a thousand charge cycles:* See this article from the battery knowledge base website, Battery University: https://batteryuniversity.com/learn/article/how_to_prolong_lithium_based_batteries.

220 *five thousand to ten thousand cycles*: Ramez Naam, "How Cheap Can Energy Storage Get? Pretty Darn Cheap," October 14, 2015. See: http://rameznaam.com/2015/10/14/how-cheap-can-energy-storage-get/.

220 *San Diego, for example:* Rob Nikolewski, "New Battery Storage Technology Connected to California Power Grid," *San Diego Union Tribune*, May 6, 2019. See: https://www.sandiegouniontribune.com/business/energy-green/story/2019-05-03/new-battery-storage-technology-connected-to-california-power-grid.

220 *Flow batteries are currently more expensive than lithium-ion batteries:* Naam, author interview.

220 *Form Energy:* Akshat Rathi, "To Hit Climate Goals, Bill Gates and His Billionaire Friends Are Betting on Energy Storage," *qz*, June 12, 2018.

220 *by as much as 50 percent:* http://spectrum.ieee.org/transportation/advanced-cars/the-charge-of-the-ultra-capacitors.

Electric Cars Are Gaining Speed

221 *one-fifth of our total energy budget:* U.S. Energy Information Administration, *Monthly Energy Review*, Table 2.1, April 2019, preliminary data. You can access the data here: https://www.eia.gov/energyexplained/use-of-energy/transportation.php.

221 *30 percent of US greenhouse gas emissions:* US Energy Information Administration, *Annual Energy Outlook 2019*, Table 36, April 2019, preliminary data. You can access the data here: https://www.eia.gov/energyexplained/use-of-energy/transportation-in-depth.php.

221 *Globally, it's a slightly smaller 20 percent:* US Energy Information Administration, *International Energy Outlook 2016*, Chapter 8, p. 131. See the report here: https://www.eia.gov/outlooks/ieo/pdf/transportation.pdf.

221 *announced the phaseout of the internal combustion engine by 2030:* (German publication) Sven Böll, "*Bundesländer wollen Benzin- und Dieselautos verbieten,*" *Spiegel Online*, October 8, 2016. See: https://www.spiegel.de/auto/aktuell/bundeslaender-wollen-benzin-und-dieselautos-ab-2030-verbieten-a-1115671.html.

221 *Norway sped past Germany: Dagens Naerlingsliv,* Tore Gjerstad, "*Frp vil fjerne bensinbilene,*" June 2, 2016. See: https://www.dn.no/motor/fremskrittspartiet/bensin/drivstoff/frp-vil-fjerne-bensinbilene/1-1-5657552.

221 *2.1 percent in the US:* The US Bureau of Transportation Statistics reports that there were approximately 272 million cars on the road in 2017. At the same time, *Scientific American* reported that there were 1 million electric cars on the road by 2018. Assuming the number of cars remained relatively constant from 2017 through 2018, electric cars made up less than 0.36 percent of cars on US roads in 2018. Maxine Joselow, "The U.S. Has 1 Million Electric Vehicles, but Does It Matter?" *Scientific American*, October 12, 2018, and bts.gov/content/number-us-aircraft-vehicles-vessels-and-other-conveyances.

221 *India is also on board:* Arkadev Ghoshal, "Watch: India Unveils Ambitious Plan to Have Only Electric Cars by 2030," *International Business Times*, April 30, 2017. See: https://www.ibtimes.co.in/watch-india-unveils-ambitious-plan-have-only-electric-cars-by-2030-724887. Also watch this presentation by Indian Minister of Railways and Commerce & Industry Piyush Goyal: https://www.youtube.com/watch?v=zCefO9qqZ_I.

221 *China's Volvo:* See this press release from Volvo: "Volvo Cars Aims for 50 Per Cent of Sales to Be Electric by 2025," April 25, 2018, https://www.media.volvocars.com/global/en-gb/media/pressreleases/227602/volvo-cars-aims-for-50-per-cent-of-sales-to-be-electric-by-2025.

221 *have all set official targets for electric car sales:* Alana Petroff, "These Countries Want to Ditch Gas and Diesel Cars," *CNN Business*, July 26, 2017. See: https://money.cnn.com/2017/07/26/autos/countries-that-are-banning-gas-cars-for-electric/index.html.

221 *$11.7 billion on ten pure electric cars and forty hybrid models*: Paul Lienert, "Global Carmakers to Invest at Least $90 Billion in Electric Vehicles," Reuters, January 15, 2018. See: https://www.reuters.com/article/us-autoshow-detroit-electric/global-carmakers-to-invest-at-least-90-billion-in-electric-vehicles-idUSKBN1F42NW.

221 *Volkswagen, dropping $40 billion to electrify:* William Boston, "VW Accelerates Electric Car Effort with $40 Billion Investment," *Wall Street Journal*, November 17, 2017.

221 *$300 billion in investments:* Paul Lienert, "Exclusive: VW, China Spearhead

$300 Billion Global Drive to Electrify Cars," Reuters, January 10, 2019. See: https://www.reuters.com/article/us-autoshow-detroit-electric-exclusive/exclusive-vw-china-spearhead-300-billion-global-drive-to-electrify-cars-idUSKCN1P40G6.

222 *Panasonic has also teamed up with Toyota:* "Toyota, Panasonic Announce Battery Venture to Expand EV Push," Reuters, January 22, 2019. See: https://www.reuters.com/article/us-toyota-panasonic/toyota-panasonic-announce-battery-venture-to-expand-ev-push-idUSKCN1PG0MP.

222 *Porsche and BMW:* See this press release about the collaboration: "Ultra-high-power Charging Technology for the Electric Vehicle of the Future," https://newsroom.porsche.com/en/company/porsche-fastcharge-prototype-charging-station-ultra-high-power-charging-technology-electric-vehicle-16606.html.

222 *Volkswagen has invested in the startup QuantumScape:* For more about QuantumScape, see: https://www.quantumscape.com/. For details on Volkswagen's $100 million transaction, see: https://www.volkswagenag.com/en/news/2018/09/QuantumScape.html. See also: Stephen Edelstein, "Volkswagen Invests $100 Million in Solid-State Battery Firm QuantumScape," *Drive*, September 16, 2018, https://www.thedrive.com/tech/23586/volkswagen-invests-100-million-in-solid-state-battery-firm-quantumscape.

222 *most electric vehicles get about 200 miles:* Electric vehicle ranges are readily available across the Web. CleanTechnica gives a good summary of the readily available data here: Loren McDonald, "US Electric Car Range Will Average 275 Miles by 2022, 400 Miles by 2028—New Research (Part 1)," *CleanTechnica,* October 27, 2018, https://cleantechnica.com/2018/10/27/us-electric-car-range-will-average-275-miles-by-2022-400-miles-by-2028-new-research-part-1/.

222 *by 15 percent a year for almost a decade:* Ibid.

222 *average mid-range model will get 275 miles*: Ibid.

222 *By 2025, the year solid state batteries are supposed to hit the market:* According to Edelstein, "Volkswagen Invests."

222 *The Tel-Aviv startup StoreDot:* See: https://www.store-dot.com/.

222 *super-capacitor:* Sometimes called ultra-capacitors, super-capacitors are a different type of energy storage device that can store a lot of energy and charge and discharge extremely quickly. For more information, see this article from Battery University: https://batteryuniversity.com/learn/article/whats_the_role_of_the_supercapacitor.

222 *five-minute charge gives you three hundred miles:* Eric Brandt, "Israeli Company Demonstrates 300-Mile Electric Car Battery That Charges in 5 Minutes," *Drive*, May 12, 2017. See: https://www.thedrive.com/news

/10227/israeli-company-demonstrates-300-mile-electric-car-battery-that-charges-in-5-minutes.

223 *150,000 gas stations in America:* For one estimate for the number of gas stations in the US, see this industry highlights page from the National Association of Convenience Stores: https://www.convenience.org/Research/FactSheets/FuelSales.

223 *sixty-eight thousand EV charging units in the US:* See the US Department of Energy's Alternative Fuel Data Center, "Electric Vehicle Charging Station Locations," September 9, 2019, https://afdc.energy.gov/fuels/electricity_locations.html#.

223 *number one charging location for EVs:* According to the US Department of Energy, US drivers do more than 80 percent of their charging at home. See: https://www.energy.gov/eere/electricvehicles/charging-home.

223 *ChargePoint:* See: https://www.chargepoint.com/. See also their commitment to build 2.5 million charging ports by 2025, here: https://www.chargepoint.com/about/news/chargepoint-makes-landmark-commitment-future-mobility-pledge-25-million-places-charge/.

223 *29.5 kilowatt-hours a day:* According to the US Energy Information Administration, the average US house consumes 867 kilowatt-hours per month, or about 29 kilowatt-hours per day. See: https://www.eia.gov/tools/faqs/faq.php?id=97&t=3.

223 *Tesla Model-S:* See: https://www.tesla.com/models.

Biodiversity and Ecosystem Services

223 *coral reefs are already in jeopardy*: Laura Parker, "Coral Reefs Could Be Gone in 30 Years," *National Geographic*, June 23, 2017. See: https://www.nationalgeographic.com/news/2017/06/coral-reef-bleaching-global-warming-unesco-sites/.

223 *25 percent of the world's biodiversity:* According to the World Wildlife Fund: https://wwf.panda.org/our_work/oceans/coasts/coral_reefs/.

224 *500 million people*: Melissa Gaskill, "The Current State of Coral Reefs," PBS, July 15, 2019. See: https://www.pbs.org/wnet/nature/blog/the-current-state-of-coral-reefs/.

224 *By 2100, a staggering 50 percent of all marine life will have disappeared:* Scott Heron, United Nations Educational, Scientific and Cultural Organization, "Impacts of Climate Change on World Heritage Coral Reefs: A First Global Scientific Assessment " World Heritage Convention, 2017.

224 *Every year, we lose 18.7 million acres of forest:* According to the World WildLife Foundation. See this website page: https://www.worldwildlife.org/threats/deforestation-and-forest-degradation.

224 *BioCarbon Engineering:* See: https://www.biocarbonengineering.com/.

225 *Dr. David Vaughan:* Sam Price-Waldman, "A Breakthrough for Coral Reef Restoration [Video]," *Atlantic*, February 22, 2016. You can watch the full story here: https://www.youtube.com/watch?v=qHKpcnn5Tws.

225 *cells allows us to grow mahi-mahi, bluefin tuna, etc.:* Clare Leschin-Hoar, "Seafood Without the Sea: Will Lab-Grown Fish Hook Consumers?" *The Salt: NPR*, May 5, 2019. See: https://www.npr.org/sections/thesalt/2019/05/05/720041152/seafood-without-the-sea-will-lab-grown-fish-hook-consumers.

225 *Harvard's E. O. Wilson:* Edward O. Wilson, *Half-Earth: Our Planet's Fight for Life*, (Livright, 2016).

225 *37 percent of the globe's landmass and 75 percent of its freshwater resources:* We draw from two World Bank datasets: (1) "Annual Freshwater Withdrawals, Agriculture (% of total freshwater withdrawal)," https://data.worldbank.org/indicator/ER.H2O.FWAG.ZS, and (2) "Agricultural Land (% of land area)," https://data.worldbank.org/indicator/AG.LND.AGRI.ZS.

226 *the* Lancet *estimates that pollution:* Philip Landriganm, "The Lancet Commission on Pollution and Health," *The Lancet Commissions* 391 (October 19, 2017): 462–512. See: https://doi.org/10.1016/S0140-6736(17)32345-0.

226 *When a cyclone hit the area in 2008, over 138,000 people died:* "Flooding and Damage from 2008 Myanmar Cyclone Assessed," *Science Daily,* August 10, 2009. See: https://www.sciencedaily.com/releases/2009/07/090717104618.htm.

227 *replanting a portion of the Irrawaddy Delta:* Adele Peters, "These Tree-Planting Drones Are About to Start an Entire Forest from the Sky,"*Fast Company,* August 10, 2017. See: https://www.fastcompany.com/40450262/these-tree-planting-drones-are-about-to-fire-a-million-seeds-to-re-grow-a-forest.

Economic Risks: The Threat of Technological Unemployment

227 *47 percent of all US jobs:* L. Nedelkoska, "Automation, Skills Use and Training," *OECD Social, Employment and Migration Working Papers*, no. 202 (OECD Publishing, Paris, 2018). See: https://doi.org/10.1787/2e2f4eea-en.

228 *as journalist and author James Surowiecki:* James Surowiecki, "Robots Will Not Take Your Job," *Wired.* August, 2017. See: https://www.wired.com/2017/08/robots-will-not-take-your-job/.

228 *In 1790, 90 percent:* See a more complete history here: https://classroom.synonym.com/during-early-1800s-americans-earned-living-what-12580.html.

228 *The agrarian economy morphed:* For more information, see the US Bureau

of Labor Statistics, "Employment by Major Industry Sector," here: https://www.bls.gov/emp/tables/employment-by-major-industry-sector.htm.

228 *Consider automatic teller machines (ATMs):* T. L. Andrews, "Robots Won't Take Your Job—They'll Help Make Room for Meaningful Work Instead," *Quartz,* March 15, 2017.

228 *T. L. Andrews pointed out in* Quartz*:* Ibid.

229 *James Wilson and Paul Daugherty in the* Harvard Business Review*:* James Wilson, "Collaborative Intelligence: Humans and AI Are Joining Forces," *Harvard Business Review,* July-August 2018. See: https://hbr.org/2018/07/collaborative-intelligence-humans-and-ai-are-joining-forces.

229 *an 85 percent increase in productivity:* Peggy Hollinger, "Meet the Cobots: Humans and Robots Together on the Factory Floor,"*National Geographic,* May 6, 2016. See: https://www.nationalgeographic.com/news/2016/05/financial-times-meet-the-cobots-humans-robots-factories/.

229 *According to research done by McKinsey:* Jame Manyika, "The Internet Created 2.6 New Jobs for Every 1 It Destroyed," McKinsey, May 2011. See: https://www.mckinsey.com/~/media/McKinsey/Industries/High%20Tech/Our%20Insights/Internet%20matters/MGI_internet_matters_exec_summary.ashx.

230 *Goldman Sachs recently made headlines:* Jon LeSage, "GOLDMAN SACHS: Self-Driving Trucks Will Kill 300,000 Jobs per Year,", *Business Insider,* November 15, 2017. See: https://www.businessinsider.com/goldman-sachs-says-self-driving-trucks-will-kill-300000-jobs-per-year-2017-11.

230 *6.7 million unfilled jobs in the US:* Jeff Cox, "The U.S. Labor Shortage Is Reaching a Critical Point," CNBC Markets, July 5, 2018. See: https://www.cnbc.com/2018/07/05/the-us-labor-shortage-is-reaching-a-critical-point.html.

Existential Risks: Vision, Prevention, and Governance

230 *Nick Bostrom:* "Existential Risks: Analyzing Human Extinction Scenarios and Related Hazards," *Journal of Evolution and Technology* 9 (March 9, 2002).

Vision

232 *Stewart Brand:* Stewart Brand, *Clock of the Long Now: Time and Responsibility, The Ideas Behind the World's Slowest Computer* (Basic Books, 1999), p.1.

Prevention

232 *Michael Kimmelman in the* New York Times*:* Michael Kimmelman, "The Dutch Have Solutions to Rising Seas. The World Is Watching," *New York Times,* June 15, 2017.

233 *"Sentry System," designed by NASA's:* Learn more about Sentry here: https://cneos.jpl.nasa.gov/sentry/vi.html.

233 *NASA's DART project:* Details about the DART Mission, otherwise known as the Double Asteroid Redirection Test, are found on NASA's website: https://www.nasa.gov/planetarydefense/dart.

233 *NASA started training AI to interpret the data:* Jackie Snow, "Future Wildfires Will Be Fought with Algorithms," *Fast Company*, November 26, 2018. See: https://www.fastcompany.com/90269483/how-ai-software-could-help-fight-future-wildfires.

Governance

234 *Baltic state of Estonia:* Nathan Heller, "Estonia, the Digital Republic," *New Yorker*, December 11, 2017. See: https://www.newyorker.com/magazine/2017/12/18/estonia-the-digital-republic.

235 *OpenGov:* See: https://opengov.com/.

235 *Social Glass:* See: https://www.social.glass/.

235 *Alphabet's Sidewalk Labs:* Alissa Walker, "Here Is Sidewalk Labs's Big Plan for Toronto," *Curbed*, June 24, 2019. See: https://www.curbed.com/2019/6/24/18715669/sidewalk-labs-toronto-alphabet-google-quayside.

Chapter Fourteen: The Five Great Migrations

237 *In their book* Exceptional People*:* Ian Goldin and Geoffrey Cameron, *Exceptional People* (Princeton University Press, 2012), p 12.

238 *Stanford economist named Petra Moser:* Clifton Parker, "Jewish Émigrés Who Fled Nazi Germany Revolutionized U.S. Science and Technology, Stanford Economist Says," *Stanford*, August 11, 2014. For an overview of this research, see: https://news.stanford.edu/news/2014/august/german-jewish-inventors-081114.html.

238 *"More than half of my colleagues at Stanford are immigrants":* Ibid.

238 *outpouring began in April of 1933:* Petra Moser, "German Jewish Émigrés and US Invention," *American Economic Review*, October, 2014.

238 *Albert Einstein and five other Nobel Laureates:* Andrew Grant, "The Scientific Exodus from Nazi Germany," *Physics Today*, September 26, 2018. See: https://physicstoday.scitation.org/do/10.1063/PT.6.4.20180926a/full/.

239 *A 2012 study by the Partnership for a New American Economy:* Partnership for a New American Economy, "Patent Pending: How Immigrants Are Reinventing the American Economy," June 2012. See: https://www.newamericaneconomy.org/sites/all/themes/pnae/patent-pending.pdf.

239 *"product reallocation":* Gaurav Khanna and Munseob Lee, "Hiring Highly

Educated Immigrants Leads to More Innovation and Better Product," *Conversation*, September 26, 2018. See: https://theconversation.com/hiring-highly-educated-immigrants-leads-to-more-innovation-and-better-products-100087.

239 *Joseph Schumpeter called "creative destruction"*: Ibid.

239 *By tracking the product reallocation rate:* Ibid.

239 *In America, immigrants are twice as likely to start a new business:* Grace Nasri, "The Shocking Stats About Who's Really Starting Companies in America," *Fast Company*, August 14, 2013. See: https://www.fastcompany.com/3015616/the-shocking-stats-about-whos-really-starting-companies-in-america.

240 *33 percent of venture-backed companies:* Mark Boslet, "NVCA Study Finds ⅓ Of Recently Public Venture Companies Have Immigrant Founders," *PE Hub Network*, June 20, 2013, See: http://nvcaccess.nvca.org/index.php/topics/public-policy/372-nvca-releases-results-from-american-made-20.html.

240 *half of all unicorns:* Stuart Anderson, "Immigrants and Billion Dollar Startups," *National Foundation for American Policy*, March 2016. See: http://nfap.com/wp-content/uploads/2016/03/Immigrants-and-Billion-Dollar-Startups.NFAP-Policy-Brief.March-2016.pdf.

Climate Migrations

241 *the very first Intergovernmental Panel on Climate Change:* "Climate Change: The IPCC 1990 and 1992 Assessments," *Intergovernmental Panel on Climate Change*, 2010. See: https://www.ipcc.ch/report/climate-change-the-ipcc-1990-and-1992-assessments/.

241 *Oxford scientist Norman Myers:* Norman Myers, "Environmental Refugees: A Growing Phenomenon of the 21st Century," *Philosophical Transactions of the Royal Society B Biological Sciences*, May 2002, DOI: 10.1098/rstb.2001.0953.

241 *as Mark Levine explained in* Outside *magazine:* Mark Levine, "A Storm at the Bone: A Personal Exploration into Deep Weather," *Outside*, November 1, 1998. See: https://www.outsideonline.com/1907231/storm-bone-personal-exploration-deep-weather.

241 *a 2015 meta-analysis of all available data:* "New Report and Maps: Rising Seas Threaten Land Home to Half a Billion," *Climate Central,* November 8, 2015. See: http://sealevel.climatecentral.org/news/global-mapping-choices.

241 *made a series of maps*: Ibid.

242 *everything south of Wall Street:* Benjamin Strauss, "American Icons Threatened by Sea Level Rise: In Pictures" *Climate Central,* October 16, 2015. See: https://www.climatecentral.org/news/american-icons-threatened

-by-sea-level-rise-in-pictures-19547#mapping-choices-us-cities-we-could-lose-to-sea-level-rise-19542.

242 *Ellie Mae O'Hagan wrote in the* Guardian*:* Ellie Mae O'Hagan, "Mass Migration Is No 'Crisis': It's the New Normal as the Climate Changes," *Guardian,* August 18, 2015. See: https://www.theguardian.com/commentisfree/2015/aug/18/mass-migration-crisis-refugees-climate-change.

242 *the 1947 partitioning of India and Pakistan:* "History's Greatest Migration," *Guardian*, September 25, 1947. See: https://www.theguardian.com/century/1940-1949/Story/0,,105131,00.html. See also: https://www.unhcr.org/3ebf9bab0.pdf.

242 *Tokyo is the largest mega-city on Earth:* Alexandre Tanzi and Wei Lu,"Tokyo's Reign as World's Largest City Fades," Bloomberg, July 13, 2018. See: https://www.bloomberg.com/news/articles/2018-07-13/tokyo-s-reign-as-world-s-largest-city-fades-demographic-trends. See also: https://en.wikipedia.org/wiki/Megacity.

Urban Relocations

243 *Three hundred years ago:* UN Population Division, *World Urbanization Prospects, the 2001 Revision* (New York, 2002).

243 *11 million Americans:* David Kennedy and Lizabeth Cohen, *The American Pageant: A History of the American People*, 15th (AP) edition (Cengage Learning, 2013), pp. 539–540.

243 *The rest of the world wasn't far behind:* Mike Davis, *Planet of Slums* (Verso, 2006).

243 *"hyper-city," a locale with a population above 20 million:* Ibid, p. 5.

243 *By 2050:* UN Population Division, *World Urbanization.*

244 *professor Richard Florida:* Richard Florida, *The New Urban Crisis* (Basic Books, 2017). See also: Richard Florida, "The Roots of the New Urban Crisis," *Citylab*, April 9, 2017, https://www.citylab.com/equity/2017/04/the-roots-of-the-new-urban-crisis/521028/.

244 *In 2016, the Brookings Institute:* Jesus Leal Trijullo and Joseph Parilla, "Redefining Global Cities: The Seven Types of Global Metro Economies," *Global Cities Initiative*, 2016. See: https://www.brookings.edu/wp-content/uploads/2016/09/metro_20160928_gcitypes.pdf.

244 *the National Bureau of Economic Research:* Edward L. Glaeser and Wentao Xiong, "Urban Productivity in the Developing World," *National Bureau of Economic Research*, March 2017. See: https://www.nber.org/papers/w23279.pdf.

244 *London and Paris:* Ibid.

244 *Santa Fe Institute physicist Geoffrey West:* Jonah Lehrer, "A Physicist Solves the City," *New York Times,* December 17, 2010. See: https://www.nytimes.com/2010/12/19/magazine/19Urban_West-t.html.

245 *2018 McKinsey study:* Katie Johnson, "Environmental Benefits of Smart City Solutions," *Foresight*, July 19, 2018. See: https://www.climateforesight.eu /cities-coasts/environmental-benefits-of-smart-city-solutions/.

Virtual Worlds

246 *three billion hours a week:* Jane McGonigal, "We Spend 3 Billion Hours a Week as a Planet Playing Videogames. Is It Worth It? How Could It Be MORE Worth It?," Ted, 2011. See: https://www.ted.com/conversations/44 /we_spend_3_billion_hours_a_wee.html.

246 *In 2005, the BBC reported:* "S Korean Dies After Games Session," *BBC News*, August 10, 2005. See: http://news.bbc.co.uk/2/hi/technology/4137782.stm.

246 *In 2014,* The Guardian: Justin McCurry, "Internet Addiction Driving South Koreans into Realms of Fantasy," *Guardian*, July 13, 2010. See: https:// www.theguardian.com/world/2010/jul/13/internet-addiction-south-korea.

246 hikikomori, *the lost generation:* Laurence Butet-Roch, "Pictures Reveal the Isolated Lives of Japan's Social Recluses," *National Geographic*, February 14, 2018. See: https://www.nationalgeographic.com/photography/proof/2018 /february/japan-hikikomori-isolation-society/.

246 *thrill ride known as dopamine:* Greg Berns, *Satisfaction* (Henry Holt and Co., 2005).

246 *Cocaine, by way of comparison:* "How Does Cocaine Produce Its Effects?" National Institute on Drug Abuse, May 2015. See: https://www.drugabuse.gov /publications/research-reports/cocaine/how-does-cocaine-produce-its-effects.

246 *dopamine dispensers dressed up like joysticks:* John Keilman, "Are Video Games Addictive like Drugs, Gambling? Some Who've Struggled Say Yes," *Chicago Tribune*, May 30, 2017. See also: Steven Kotler, *The Rise of Superman* (New Harvest, 2013), p. 98.

247 *Nearly all the major uses of the internet:* Trevor Haynes, "Dopamine, Smartphones & You: A Battle for Your Time," Harvard University: the Graduate School for Arts & Sciences, May 1, 2018. See: http://sitn.hms .harvard.edu/flash/2018/dopamine-smartphones-battle-time/.

247 *virtual environment spikes dopamine:* Steven Kotler, "Legal Heroin: Is Virtual Reality Our Next Hard Drug," *Forbes*, January 15, 2014. See: https://www .forbes.com/sites/stevenkotler/2014/01/15/legal-heroin-is-virtual-reality-our -next-hard-drug/#1cb0e6511a01.

247 *psychiatrist Keith Ablow:* "Facebook Twists Reality Again and Risks Ruining Your Children," Fox News, May 3, 2014. See: https://www.foxnews.com /opinion/facebook-twists-reality-again-and-risks-ruining-your-children.

247 *VR makes it able to trigger all six:* Kotler, "Legal Heroin."

247 *research into flow states:* Kotler, *Rise of Superman.*

248 *Back in 2006,* BusinessWeek*:* "My Virtual Life," *Bloomberg Businessweek*,

April 30, 2006. See: https://www.bloomberg.com/news/articles/2006-04-30/my-virtual-life.

248 *bestselling novel* Ready Player One*:* Ernest Cline, *Ready Player One* (Broadway Books, 2012).

249 *Al Cooper:* Al Cooper, "Online Sexual Compulsivity: Getting Tangled in the Net," *Sex Addict Compulsive*, 1999, pp. 79–104. See also: D. Damania, "Internet Pornography Statistics," *d infographics*, December 23, 2011, http://thedinfographics.com/2011/12/23/internet-pornography-statistics/.

249 *321 million Americans already spend eleven hours:* "The Nielsen Total Audience Report: Q1 2018," The Nielsen Company, 2018. See: http://www.nielsen.com/us/en/insights/reports/2018/q1-2018-total-audience-report.html.

Space Migration

249 *Konstantin Tsiolkovsky:* Dennis Overbye, "Is It Time to Play with Space Again?," *New York Times*, July 15, 2019. See: https://www.nytimes.com/2019/07/15/science/apollo-moon-space.html.

249 *Tsiolkovsky was a true visionary:* "Konstantin E. Tsiolkovsky," NASA, September 22, 2010. See: https://www.nasa.gov/audience/foreducators/rocketry/home/konstantin-tsiolkovsky.html.

250 *"back up the biosphere":* John Gilbey, "Backing Up the Biosphere," *Nature*, April 7, 2012. See: https://www.nature.com/news/backing-up-the-biosphere-1.10395.

250 *@JeffBezos:* See: ttps://twitter.com/JeffBezos.

250 *@elonmusk:* See: ttps://twitter.com/ElonMusk.

250 *We'll start with Bezos:* Jeff Bezos, author interview, 2015. See also: Catherine Clifford, "Jeff Bezos: You Can't Pick Your Passions," *CNBC Make It*, February 7, 2019, https://www.cnbc.com/2019/02/07/amazon-and-blue-origins-jeff-bezos-on-identifying-your-passion.html.

251 *"a simple two-step plan":* Tony Reichhardt, "Jeff Bezos' Simple Two-Step Plan," *Air & Space*, September 16, 2016. See: https://www.airspacemag.com/daily-planet/jeff-bezos-simple-two-step-plan-180960498/.

251 *the Moon—which he still believes is the best launch spot:* Korey Haynes, "O'Neill Colonies: A Decades-Long Dream for Settling Space," *Astronomy*, May 17, 2019. See: http://www.astronomy.com/news/2019/05/oneill-colonies-a-decades-long-dream-for-settling-space.

251 *Bezos at a 2019 event in Washington, DC:* Neel V. Patel, "Jeff Bezos Wants to Solve All Our Problems by Shipping Us to the Moon," *Popular Science*, May 9, 2019. See: https://www.popsci.com/blue-origin-moon-lander/.

251 *Blue Moon Lunar Lander:* Ibid. See also: Ian Allen, "Jeff Bezos Wants Us All to Leave Earth—for Good," *Wired*, October 15, 2018, https://www.wired.com/story/jeff-bezos-blue-origin/.

251 *"The Earth is the gem":* Loren Grush, "Jeff Bezos: 'I Don't Want a Plan B for Earth,'" *Verge*, June 1, 2016. See: https://www.theverge.com/2016/6/1/11830206/jeff-bezos-blue-origin-save-earth-code-conference-interview.

251 *Elon Musk doesn't disagree:* Dave Mosher, "Here's Elon Musk's Complete, Sweeping Vision on Colonizing Mars to Save Humanity," *Business Insider*, September 29, 2016. See: https://www.businessinsider.com/elon-musk-mars-speech-transcript-2016-9.

252 *"Mars Oasis":* Chris Anderson, "Elon Musk's Mission to Mars," *Wired*, September 21, 2012. See: https://www.wired.com/2012/10/ff-elon-musk-qa/.

252 *Musk founded SpaceX in 2002:* Michael Sheetz, "The Rise of Spacex and the Future Of Elon Musk's Mars Dream," CNBC, March 20, 2019. See: https://www.cnbc.com/2019/03/20/spacex-rise-elon-musk-mars-dream.html. See also: Tim Fernholz, "The Complete Visual History of Spacex's Single-Minded Pursuit of Rocket Reusability," *Quartz*, July 1, 2017, https://qz.com/1016072/a-multimedia-history-of-every-single-one-of-spacexs-attempts-to-land-its-booster-rocket-back-on-earth/.

253 *began work on "Starship":* Mike Wall, "Big Leap by SpaceX's Starship Prototype Pushed to Next Week," *Space*, August 16, 2019. See: https://www.space.com/spacex-starhopper-big-test-flight-target-date.html.

253 *a problem to be solved this decade:* Matt Williams, "Musk Gives an Update on When a Mars Colony Could Be Built," *Universe Today*, September 25, 2018. See: https://www.universetoday.com/140071/musk-gives-an-update-on-when-a-mars-colony-could-be-built/.

253 *Musk thinks about $500,000:* Amanda Kooser, "Elon Musk Expects Spacex Ticket to Mars Will Cost $500,000," CNET, February 11, 2019. See: https://www.cnet.com/news/elon-musk-expects-spacex-ticket-to-mars-will-cost-500000/.

Meta-Intelligence: Into the Borg

253 *In 2015, Harvard chemist Charles Lieber:* Jung Min Lee, "Nanoenabled Direct Contact Interfacing of Syringe-Injectable Mesh Electronics," Nano Letters, 2019. See: http://cml.harvard.edu/assets/Nanoenabled-Direct-Contact-Interfacing-of-Syringe-Injectable-Mesh-Electronics.pdf.

254 *Polina Anikeeva pointed out in her 2015 TED Talk:* See: https://www.youtube.com/watch?v=MZ3Q638aMlA.

255 *Lieber injected the mesh into the retina of a mouse:* Guosong Hong, "A Method for Single Neuron Chronic Recording from the Retina in Awake Mice," *Science*, June 29, 2018. See: https://www.ncbi.nlm.nih.gov/pmc/articles/PMC6047945/.

255 *"a neural lace," an injectable brain-computer interface:* Eric Lutz, "Elon Musk Has Created "Threads" to Weave a Computer into Your Brain," *Vanity Fair*,

July 17, 2019. See: https://www.vanityfair.com/news/2019/07/elon-musk-neuralink-created-threads-to-weave-computer-into-your-brain.

256 *BCIs have allowed paraplegics to walk again:* Laura Kauhanen, "EEG-Based Brain-Computer Interface for Tetraplegics," *Computational Intelligence and Neuroscience*, September 19, 2007. See: https://www.ncbi.nlm.nih.gov/pmc/articles/PMC2233767/.

256 *Stroke victims, paralyzed:* "Brain-Computer Interface Enables Paralyzed Man to Walk Without Robotic Support," *Kurzweil*, September 25, 2015. See: https://www.kurzweilai.net/brain-computer-interface-enables-paralyzed-man-to-walk-without-robotic-support.

256 *Epileptics have been cured of seizures:* Rafeed Alkawadri, "Brain–Computer Interface (BCI) Applications in Mapping of Epileptic Brain Networks Based On Intracranial-EEG: An Update," *Frontiers in Neuroscience*, March 27, 2019. See: https://www.frontiersin.org/articles/10.3389/fnins.2019.00191/full.

256 *a team of Harvard researchers:* Linda Xu, "Humans, Computers and Everything In Between: Towards Synthetic Telepathy," *Harvard Science Review*, May 1, 2014. See: https://harvardsciencereview.com/2014/05/01/synthetic-telepathy/.

256 *EEG headsets to play video games:* Bojan Kerous, "EEG-Based BCI and Video Games: A Progress Report," *Virtual Reality*, June 2018, pp. 119–135.

257 *Loneliness, according to too many studies:* Natalie Gil, "Loneliness: A Silent Plague That Is Hurting Young People Most," *Guardian*, July 20, 2014. See: https://www.theguardian.com/lifeandstyle/2014/jul/20/loneliness-britains-silent-plague-hurts-young-people-most.

257 *"group flow":* Keith Sawyer, *Group Genius* (Basic Books, 2017).

258 *the trajectory of evolution:* See for example: Larry S. Yaeger, "How Evolution Guides Complexity," *HFSP Journal*, October 2009, https://www.ncbi.nlm.nih.gov/pmc/articles/PMC2801533/. See also: Brandon Keim, "The Complexity of Evolution," *Wired*, April 15, 2008, https://www.wired.com/2008/04/the-complexity/.

258 *Neuralink has a plan:* Elizabeth Lopatto, "Elon Musk Unveils Neuralink's Plans for Brain-Reading 'Threads' and a Robot to Insert Them," *Verge*, July 16, 2019. See: https://www.theverge.com/2019/7/16/20697123/elon-musk-neuralink-brain-reading-thread-robot.

INDEX

ABOUT THE AUTHORS

Peter H. Diamandis is a *New York Times* bestselling author and the founder of more than fifteen high-tech companies. He is the CEO of the XPRIZE and executive chairman of Singularity University, a Silicon Valley–based institution backed by Google, 3-D Systems, and NASA. He is cochairman of Planetary Resources, Inc., and the cofounder of Human Longevity, Inc. Diamandis attended MIT, where he received his degrees in molecular genetics and aerospace engineering, and Harvard Medical School, where he received his MD. In 2014 he was named one of "The World's 50 Greatest Leaders" by *Fortune* magazine.

Steven Kotler is a *New York Times* bestselling author, award-winning journalist, and the founder and executive director of the Flow Research Collective. He is considered one of the world's leading experts in peak performance. His books include *Stealing Fire*, *BOLD*, *The Rise of Superman*, *Abundance*, *A Small Furry Prayer*, *West of Jesus*, and *Last Tango in Cyberspace*. His work has been nominated for two Pulitzer Prizes, has been translated into more than forty languages, and has appeared in over a hundred publications, including the *New York Times Magazine*, *Atlantic Monthly*, *Wired*, *Forbes*, and *Time*.